Cyclic Designs

J.A. JOHN

Department of Mathematics
University of Southampton

Springer-Science+Business Media, B.V.

Originally published by Chapman and Hall in 1987.
Softcover reprint of the hardcover 1st edition 1987
ISBN 978-0-412-28240-9 ISBN 978-1-4899-3326-3 (eBook)
DOI 10.1007/978-1-4899-3326-3

British Library Cataloguing in Publication Data

John, J.A.
Cyclic designs. —— (Monographs on statistics and applied probability)
1. Experimental design
I. Title II. Series
001.4'34'028 QA279

Library of Congress Cataloging in Publication Data

John, J.A.
Cyclic designs.
(Monographs on statistics and applied probability)
Bibliography: p
Includes index.
1. Experimental design. 2. Incomplete block designs.
I. Title. II. Series.
QA279.J63 1987 519.5 86–12964

TM

MONOGRAPHS ON
STATISTICS AND APPLIED PROBABILITY

General Editors

D.R. Cox, D.V. Hinkley, D. Rubin and B.W. Silverman

The Statistical Analysis of Compositional Data
J. Aitchison

Probability, Statistics and Time
M.S. Bartlett

The Statistical Analysis of Spatial Pattern
M.S. Bartlett

Stochastic Population Models in Ecology and Epidemiology
M.S. Bartlett

Risk Theory
R.E. Beard, T. Pentikäinen and E. Pesonen

Bandit Problems Sequential Allocation of Experiments
D.A. Berry and B. Fristedt

Residuals and Influence in Regression
R.D. Cook and S. Weisberg

Point Processes
D.R. Cox and V. Isham

Analysis of Binary Data
D.R. Cox

The Statistical Analysis of Series of Events
D.R. Cox and P.A.W. Lewis

Analysis of Survival Data
D.R. Cox and D. Oakes

Queues
D.R. Cox and W.L. Smith

Stochastic Modelling and Control
M.H.A. Davis and R. Vinter

Stochastic Abundance Models
S. Engen

The Analysis of Contingency Tables
B.S. Everitt

An Introduction to Latent Variable Models
B.S. Everitt

Finite Mixture Distributions
B.S. Everitt and D.J. Hand

Population Genetics
W.J. Ewens

Classification
A.D. Gordon

Monte Carlo Methods
J.M. Hammersley and D.C. Handscomb

Identification of Outliers
D.M. Hawkins

Generalized Linear Models
P. McCullagh and J.A. Nelder

Distribution-free Statistical Methods
J.S. Maritz

Multivariate Analysis in Behavioural Research
A.E. Maxwell

Applications of Queueing Theory
G.F. Newell

Some Basic Theory for Statistical Inference
E.J.G. Pitman

Density Estimation for Statistics and Data Analysis
B.W. Silverman

Statistical Inference
S.D. Silvey

Models in Regression and Related Topics
P. Sprent

Regression Analysis with Applications
G.B. Wetherill

Sequential Methods in Statistics
G.B. Wetherill and K.D. Glazebrook

(Full details concerning this series are available from the Publishers)

Contents

Preface

The biggest single influence on the development of the subject of *design of experiments* over the past quarter-century has been the availability of computers. Prior to the computer it was essential that any design had a straightforward method of analysis, which meant that the mathematical and combinatorial properties of the designs were of primary importance. Many of the designs proposed and studied also possessed important statistical properties and thus continue to be practically useful but, now that ease of analysis is less important, a very large number of these designs no longer have any real value. With the advent of the computer it has become possible to study families of designs which have relatively simple methods of construction and which provide large numbers of designs. Within a family, the designs which satisfy certain desirable statistical, rather than mathematical, properties can then be identified using a combination of theory and computing. One of the primary aims of this monograph is to study families of block and row–column designs, both unifactor and multifactor, whose methods of construction are cyclical in nature; hence the title *Cyclic Designs.*

The usual practice adopted in books on the design of experiments is to follow the description of a particular design by its method of analysis and, possibly, a numerical example. While there is still a place for such an approach in introductory texts, the ease and speed with which matrix calculations can now be carried out on computers means that, in practice, separate analyses for different classes of designs are no longer necessary. It is possible for all block designs or all row–column designs, whether unifactor or multifactor, to be readily analysed using a single computer program. All the designs in this monograph can be analysed using the general approach given in Chapter 8. Only two numerical examples are included: one in Chapter 8 to illustrate the method of analysis for block designs, and one in Chapter 7 to show the additional computations required to analyse a factorial experiment. Of course the properties of a design are closely linked to its analysis so that Chapters 1 and 2 develop the theory and criteria needed to judge and compare different designs of the same size. The remaining chapters are concerned with unifactor and multifactor block and row–column designs.

A number of other topics, such as changeover designs and split-plot designs, have been omitted mainly because cyclical methods of constructing such designs have yet to be fully developed. Any treatment given in this book would thus simply duplicate material already available in other sources.

The book is aimed at the postgraduate level and will hopefully be of interest to students, to those planning and designing experiments and to those engaged in research in the subject. It is assumed that the reader is familiar with the basic principles involved in experimentation, with the requirements for a good experiment, with the concept of factorial experimentation and with some of the standard designs, such as randomized blocks, Latin squares and balanced incomplete blocks. Such material is covered in most books on the design of experiments; see, for example, Cox (1958), Fisher (1960) and John and Quenouille (1977). Matrix algebra is used extensively in the monograph. The main results required are given in the Appendix; the reader's attention is particularly drawn to the definition of the important matrix **K** in Section A.1.

The structure of the monograph developed as a result of many stimulating discussions with Professor H.D. Patterson, formerly of the AFRC Unit of Statistics at the University of Edinburgh. I am deeply indebted to him for his advice, friendship and constant encouragement. I am also very grateful to Dr. J.A. Eccleston of the University of New South Wales who read an earlier draft and made a number of useful suggestions and criticisms. Finally, I would like to thank Christine Best at the AFRC Unit of Statistics, University of Edinburgh, for the speed and accuracy with which she typed the manuscript.

Southampton J.A. John
1986

CHAPTER 1

Analysis of block designs

1.1 Introduction

The foundations of the statistical approach to experimentation were laid by R.A. Fisher in the early 1930s. The subject evolved in agriculture but is now applicable to almost all sciences, to engineering and to some arts. The aim of an experiment is to compare a number of treatments on the basis of the responses produced in the experimental material. In agriculture the treatments may be varieties of wheat or different fertilizers; in engineering they may be temperature levels; and in a chemical experiment the purpose may be to compare several catalysts. The confidence and accuracy with which treatment differences can be assessed will depend to a large extent on the size of the experiment and on the inherent variability in the experimental material. Dividing the material into relatively homogeneous blocks can be an important technique for improving the precision of an experiment.

Suppose that n experimental units or *plots* are available for experimentation and that they are divided into b *blocks* such that the jth block consists of k_j plots ($j = 1, 2, \ldots, b$). The total variation in the data, based on $n - 1$ degrees of freedom, can then be partitioned into two components: one measuring the variation *between blocks* and the other measuring the variation between plots *within blocks*, based on, respectively, $b - 1$ and $\sum(k_j - 1) = n - b$ degrees of freedom. This partition of the degrees of freedom is shown in Table 1.1.

Since blocks are chosen so that the material within them is relatively homogeneous, comparisons made within blocks are usually more accurate than comparisons between blocks. The allocation of treatments to plots should, therefore, be carried out in such a way that important treatment comparisons are made within blocks. The actual allocation of the treatments to plots is called the *design* of the experiment. In some designs it will be possible for all treatment

Table 1.1 *Partition of degrees of freedom in a blocked experiment.*

	Degrees of freedom
Between blocks	$b-1$
Within blocks (between plots)	$n-b$
Total	$n-1$

comparisons to be estimated within blocks. In other cases, restrictions on the experiment, imposed perhaps by the shortage of suitable experimental material or by a large number of treatments under test, may make this impossible. In such cases it may be necessary to estimate certain comparisons between blocks; treatments should then be allocated wherever possible to ensure that these comparisons are the ones of least importance. In other situations it may be necessary to obtain the information on treatment comparisons in part from both within and between blocks; again the aim should be to obtain as much information as possible on important comparisons within blocks.

In the first instance, experiments involving the allocation of a *single set* of treatments to plots will be considered. The treatment comparisons of interest will depend on the composition of this single set and on the purpose of the experiment. For example, a nutritional experiment may be concerned with the effects of a number of diets using litters of mice. The litters constitute blocks since large differences between litters can often be expected. These litter effects are eliminated in any within-litter comparison so that dietary comparisons made within litters are likely to be more accurate than those made between litters. With a completely distinct, unstructured set of diets, each diet would probably be compared with every other diet. However, other comparisons may be of interest if there is some structure. If one diet is primarily based on fish while the other diets are free of fish, then a comparison of the fish diet with the others may be of special interest.

For a slightly more complex example, consider a subjective aircraft noise experiment where subjects (blocks) are asked to assess the noisiness of four different types of aircraft, namely a supersonic jet, a subsonic jumbo jet and two subsonic smaller jets. Interest may now centre around comparing the supersonic jet with the three subsonic jets and in comparing the jumbo jet with the two smaller jets. For quantitative treatments, a different set of comparisons may be of importance. Suppose a chemical experiment is designed to study the effect of different temperature levels on a certain process. Now it may be of interest to see how yield varied as temperature level increased; whether the change was linear or non-linear. Temperature comparisons which measure linearity or some form of non-linearity would then be examined.

The above are examples of experiments involving single sets of treatments, even though not all treatment comparisons may be of equal interest. More complex treatment structures arise when the experiments involve more than one set of treatments. For example, in a fertilizer experiment the treatments may involve a combination of different amounts of nitrate and potash. The effects of differing amounts of nitrate and potash could be examined separately. In addition, however, the effectiveness of nitrate at different levels of potash could be studied, and vice versa. Such treatment structures give rise to different design problems and will be considered in detail in Chapters 6 and 7.

The analysis of an experiment in which treatment comparisons are estimated within blocks only is called an *intra-block* analysis, and will be considered in this chapter. The allocation of treatments to blocks in some optimal way will be the primary aim of this book. Optimality criteria can be based on the precision with which estimates of treatment comparisons are made in the intra-block analysis; these criteria are discussed in Chapter 2. If the blocks can be regarded as a random sample of blocks from some population, then estimates of treatment comparisons may also be available from between-block differences, giving rise to an *inter-block* analysis. Where information is available from both between and within blocks, the intra- and inter-block estimates can be *combined* to provide overall estimates of treatment comparisons. The problem of how this can be done for block designs, and other blocking structures, is considered in Chapter 8.

Before applying any of the designs obtained by the methods given in this book, they will have to be randomized. Randomization ensures the validity of the experiment in the sense that the conclusions are free from the biases of the experimenter. Although randomization is briefly discussed in Chapter 8, a fuller account of the need for randomization and of how different designs are randomized can be found in most of the introductory texts on the design and analysis of experiments. In particular, excellent accounts can be found in Fisher (1960) and Cox (1958).

1.2 Intra-block model

A single set of v treatments is applied to n experimental plots which have been divided into b blocks with k_j plots in the jth block ($j = 1, 2, \ldots, b$). The expected value of the response y_{ijk} obtained when the ith treatment is applied in the jth block on the kth occasion will be assumed to be the sum of three separate components, a general mean parameter μ, a parameter τ_i measuring the effect of the ith treatment and a parameter β_j giving the effect of the jth block. It will be assumed that the responses are uncorrelated, with the same variance σ^2.

If n_{ij} is the number of times that the ith treatment is applied to the jth block, then the model can be written as

$$y_{ijk} = \mu + \tau_i + \beta_j + \varepsilon_{ijk} \tag{1.1}$$
$$(i = 1, 2, \ldots, v; j = 1, 2, \ldots, b; k = 1, 2, \ldots, n_{ij})$$

where the error terms ε_{ijk} are uncorrelated random variables each with mean zero and variance σ^2. In the intra-block analysis the block effects are taken to be a fixed set of parameters, and are not to be regarded as random variables. One of the primary objectives of the analysis will be to obtain estimates of the treatment parameters τ_i, together with the standard errors of these estimates. It may also be important to establish whether the treatment comparisons of interest differ significantly from zero. The additive model (1.1) will be assumed to be an appropriate model for block designs. In practice, of course, it is important that the assumptions underlying this model are carefully examined. Much useful information can be obtained by studying the differences between the observed responses and those obtained from the fitted model, i.e. from an analysis of the residuals. Details of such analyses can be found in many statistical texts, e.g. Draper and Smith (1981) and John and Quenouille (1977).

In matrix notation, model (1.1) can be written as

$$\mathbf{y} = \mathbf{W}\boldsymbol{\alpha} + \boldsymbol{\varepsilon} \tag{1.2}$$

where $\mathbf{y}$ and $\boldsymbol{\varepsilon}$ are $n \times 1$ vectors of response and error terms respectively, $\boldsymbol{\alpha}$ is a $p \times 1$ vector of parameters, and $\mathbf{W}$ an $n \times p$ design matrix of ones and zeros, and where $p = 1 + v + b$. It is assumed that $E(\boldsymbol{\varepsilon})$, the expected value of the error vector, is zero and that $V(\boldsymbol{\varepsilon})$, the variance–covariance matrix of the error vector, is $\sigma^2\mathbf{I}$.

Many of the features of this model and of the subsequent algebra can be illuminated by following through the various stages with a hypothetical example of a small block design. The example is used for illustrative purposes only and it is not necessarily suggested that such a design be used in practice.

Example 1.1

An experiment is carried out with three treatments set out in four blocks. The responses obtained are shown in Table 1.2.

Letting τ_1, τ_2 and τ_3 be the treatment parameters corresponding to treatments A, B and C respectively, the model assumed for this experiment is, using the matrix form given in (1.2), as follows:

$$\begin{pmatrix} 8 \\ 14 \\ 9 \\ 5 \\ 13 \\ 16 \\ 12 \\ 9 \\ 8 \end{pmatrix} = \left(\begin{array}{c|ccc|cccc} 1 & 1 & 0 & 0 & 1 & 0 & 0 & 0 \\ 1 & 0 & 0 & 1 & 1 & 0 & 0 & 0 \\ 1 & 0 & 1 & 0 & 0 & 1 & 0 & 0 \\ 1 & 0 & 0 & 1 & 0 & 1 & 0 & 0 \\ 1 & 1 & 0 & 0 & 0 & 0 & 1 & 0 \\ 1 & 0 & 1 & 0 & 0 & 0 & 1 & 0 \\ 1 & 0 & 0 & 1 & 0 & 0 & 1 & 0 \\ 1 & 1 & 0 & 0 & 0 & 0 & 0 & 1 \\ 1 & 0 & 1 & 0 & 0 & 0 & 0 & 1 \end{array}\right) \begin{pmatrix} \mu \\ \hline \tau_1 \\ \tau_2 \\ \tau_3 \\ \hline \beta_1 \\ \beta_2 \\ \beta_3 \\ \beta_4 \end{pmatrix} + \begin{pmatrix} \varepsilon_1 \\ \varepsilon_2 \\ \varepsilon_3 \\ \varepsilon_4 \\ \varepsilon_5 \\ \varepsilon_6 \\ \varepsilon_7 \\ \varepsilon_8 \\ \varepsilon_9 \end{pmatrix}$$

Note that the columns of the $\mathbf{W}$ matrix can be conveniently partitioned as indicated into three submatrices corresponding

Table 1.2 *Responses for hypothetical experiment.*

	Treatment		
Block	*A*	*B*	*C*
1	8		14
2		9	5
3	13	16	12
4	9	8	

the parameter μ, the treatment parameters τ_1, τ_2 and τ_3, and the block parameters $\beta_1, \beta_2, \beta_3$ and β_4.

Hence, the matrix $\mathbf{W}$ and the vector $\boldsymbol{\alpha}$ in (1.2) can be partitioned as follows:

$$\mathbf{W} = (\mathbf{1} \mid \mathbf{X} \mid \mathbf{Z})$$
$$\boldsymbol{\alpha} = (\mu \mid \boldsymbol{\tau}' \mid \boldsymbol{\beta}')'$$

where $\mathbf{1}$ is a column vector of ones and where $\mathbf{X}$ and $\mathbf{Z}$ are the $n \times v$ and $n \times b$ *design* matrices for treatments and blocks respectively. The model (1.2) can now be rewritten as

$$\mathbf{y} = \mathbf{1}\mu + \mathbf{X}\boldsymbol{\tau} + \mathbf{Z}\boldsymbol{\beta} + \boldsymbol{\varepsilon} \tag{1.3}$$

Some further notation will be introduced to simplify the algebra of subsequent sections. In the treatment design matrix $\mathbf{X}$ there is a single column for each treatment. The sum (and sum of squares) of the elements in any column gives the number of times the corresponding treatment occurs in the design. Further, the sum of cross-products of any two columns must be zero since any plot can only contain one treatment at a time. Hence, if $\mathbf{r}$ is the *replication vector*, whose elements give the number of times each treatment occurs in the design, then

$$\mathbf{X}'\mathbf{1} = \mathbf{r} \qquad \text{and} \qquad \mathbf{X}'\mathbf{X} = \mathbf{r}^{\delta} \tag{1.4}$$

where $\mathbf{r}^{\delta}$ is the diagonal matrix whose diagonal elements are those of the vector $\mathbf{r}$.

Similarly, if $\mathbf{k}$ is a vector whose jth element is k_j, the number of plots in the jth block, then

$$\mathbf{Z}'\mathbf{1} = \mathbf{k} \qquad \text{and} \qquad \mathbf{Z}'\mathbf{Z} = \mathbf{k}^{\delta}. \tag{1.5}$$

It is readily verified in Example 1.1 that these calculations give $\mathbf{r} = (3\ \ 3\ \ 3)'$ and $\mathbf{k} = (2\ \ 2\ \ 3\ \ 2)'$.

The sum of cross-products of the ith column of $\mathbf{X}$ and the jth column of $\mathbf{Z}$ will give the number of times the ith treatment occurs in the jth block, namely n_{ij}. If $\mathbf{N}$ is the matrix whose (ij)th element is n_{ij}, then

$$\mathbf{X}'\mathbf{Z} = \mathbf{N} \tag{1.6}$$

The $v \times b$ matrix $\mathbf{N}$, which is called the *incidence matrix* of the design, has a row for each treatment and a column for each block. For

Example 1.1,

$$\mathbf{N} = \begin{pmatrix} 1 & 0 & 1 & 1 \\ 0 & 1 & 1 & 1 \\ 1 & 1 & 1 & 0 \end{pmatrix}$$

Finally, if **T** and **B** are vectors of treatment and block totals respectively and if G is the overall total, then from a vector multiplication of the columns of **W** with the response vector **y** it is established that

$$\mathbf{X}'\mathbf{y} = \mathbf{T}, \qquad \mathbf{Z}'\mathbf{y} = \mathbf{B}, \qquad \mathbf{1}'\mathbf{y} = G \tag{1.7}$$

Again for Example 1.1, $\mathbf{T} = (30\ \ 33\ \ 31)'$, $\mathbf{B} = (22\ \ 14\ \ 41\ \ 17)'$ and $G = 94$.

1.3 Least squares analysis

The least squares estimator of the parameter vector $\boldsymbol{\alpha}$ in (1.2) is obtained by minimizing the error sum of squares $\boldsymbol{\varepsilon}'\boldsymbol{\varepsilon}$ with respect to $\boldsymbol{\alpha}$. Since

$$\boldsymbol{\varepsilon}'\boldsymbol{\varepsilon} = (\mathbf{y} - \mathbf{W}\boldsymbol{\alpha})'(\mathbf{y} - \mathbf{W}\boldsymbol{\alpha}) = \mathbf{y}'\mathbf{y} - 2\mathbf{y}'\mathbf{W}\boldsymbol{\alpha} + \boldsymbol{\alpha}'\mathbf{W}'\mathbf{W}\boldsymbol{\alpha}$$

then

$$\partial(\boldsymbol{\varepsilon}'\boldsymbol{\varepsilon})/\partial\boldsymbol{\alpha} = -2\mathbf{W}'\mathbf{y} + 2\mathbf{W}'\mathbf{W}\boldsymbol{\alpha}$$

Equating these partial derivatives to zero gives the *normal equations* $\mathbf{W}'\mathbf{W}\hat{\boldsymbol{\alpha}} = \mathbf{W}'\mathbf{y}$ with a solution of these equations producing an estimate $\hat{\boldsymbol{\alpha}}$ of the true parameter vector $\boldsymbol{\alpha}$. Partitioning **W** and $\boldsymbol{\alpha}$ as in the previous section gives the normal equations, for the model (1.3), as

$$n\hat{\mu} + \mathbf{r}'\hat{\boldsymbol{\tau}} + \mathbf{k}'\hat{\boldsymbol{\beta}} = G \tag{1.8}$$

$$\mathbf{r}\hat{\mu} + \mathbf{r}^{\delta}\hat{\boldsymbol{\tau}} + \mathbf{N}\hat{\boldsymbol{\beta}} = \mathbf{T} \tag{1.9}$$

$$\mathbf{k}\hat{\mu} + \mathbf{N}'\hat{\boldsymbol{\tau}} + \mathbf{k}^{\delta}\hat{\boldsymbol{\beta}} = \mathbf{B} \tag{1.10}$$

where use has been made of (1.4)–(1.7).

These equations do not have an unique solution since they are less than full rank. This can be seen by noting that the sum of the v rows of (1.9) and the sum of the b rows of (1.10) both give equation (1.8). Thus, the rank of the matrix of coefficients $\mathbf{W}'\mathbf{W}$ is at least two less than the size of the matrix, i.e. rank $(\mathbf{W}'\mathbf{W}) \leqslant v + b - 1$.

A solution, or solutions, to the normal equations can, however, be readily found. One method, for instance, is to put a set of parameters equal to zero, or to other arbitrary values, in such a way that the remaining equations are of full rank. In general, any solution to the normal equations $\mathbf{W'W}\hat{\boldsymbol{\alpha}} = \mathbf{W'y}$ can be written as $\hat{\boldsymbol{\alpha}} = (\mathbf{W'W})^{-}\mathbf{W'y}$, where $(\mathbf{W'W})^{-}$ is a generalized inverse of $\mathbf{W'W}$, satisfying $\mathbf{W'W}(\mathbf{W'W})^{-}\mathbf{W'W} = \mathbf{W'W}$. A brief discussion of generalized inverses is given in the Appendix.

The task of finding a solution in the present case is simplified if a set of equations involving only the treatment parameters (the parameters of interest) is obtained by eliminating $\hat{\mu}$ and $\hat{\boldsymbol{\beta}}$ from the normal equations. This is achieved by first premultiplying (1.10) by $\mathbf{Nk}^{-\delta}$, where $\mathbf{k}^{-\delta}$ is the inverse of the matrix $\mathbf{k}^{\delta}$, to give

$$\mathbf{r}\hat{\mu} + \mathbf{Nk}^{-\delta}\mathbf{N}'\hat{\boldsymbol{\tau}} + \mathbf{N}\hat{\boldsymbol{\beta}} = \mathbf{Nk}^{-\delta}\mathbf{B}$$

and then subtracting this equation from (1.9) to give

$$\mathbf{A}\hat{\boldsymbol{\tau}} = \mathbf{q} \tag{1.11}$$

where

$$\mathbf{A} = \mathbf{r}^{\delta} - \mathbf{Nk}^{-\delta}\mathbf{N}' \tag{1.12}$$

and

$$\mathbf{q} = \mathbf{T} - \mathbf{Nk}^{-\delta}\mathbf{B}. \tag{1.13}$$

A solution to the normal equations can now be obtained by solving the equations (1.11) for $\hat{\boldsymbol{\tau}}$ and then from (1.10), obtaining $\hat{\boldsymbol{\beta}}$ as

$$\hat{\boldsymbol{\beta}} = \mathbf{k}^{-\delta}\mathbf{B} - \mathbf{k}^{-\delta}\mathbf{N}'\hat{\boldsymbol{\tau}} - \mathbf{1}\hat{\mu}. \tag{1.14}$$

An intuitively appealing estimate of μ is given by $\hat{\mu} = \bar{y}$, although any other value can be used.

Hence the problem of finding a solution to the normal equations has now been simplified to one of finding a solution to (1.11).

The equations (1.11) are called the *reduced normal equations* for the treatment parameters obtained by eliminating the mean and block parameters from the full set of normal equations. The $v \times v$ coefficient matrix $\mathbf{A}$ is called the *information matrix* of the design in the intra-block analysis and plays a key role, in particular in establishing suitable optimality criteria for choosing between different block designs, as will be shown in the next chapter. The vector $\mathbf{q}$ is called the vector of *adjusted treatment totals* since linear combinations of the block totals $\mathbf{B}$ are subtracted from the treatment totals **T**. Note that the vector involves linear combinations of the responses since it can be written as

$$\mathbf{q} = (\mathbf{X}' - \mathbf{N}\mathbf{k}^{-\delta}\mathbf{Z}')\mathbf{y} = \mathbf{X}'[\mathbf{I} - \mathbf{Z}(\mathbf{Z}'\mathbf{Z})^{-1}\mathbf{Z}']\mathbf{y}$$

Again no unique solution of the reduced normal equations exists since the information matrix **A** is not of full rank. The rows and columns of **A** sum to zero, so that $\text{rank}(\mathbf{A}) \leqslant v - 1$. All solutions to the equations (1.11) can be written as

$$\hat{\boldsymbol{\tau}} = \boldsymbol{\Omega}\mathbf{q} \qquad (1.15)$$

where $\boldsymbol{\Omega}$ is a generalized inverse of **A** satisfying $\mathbf{A}\boldsymbol{\Omega}\mathbf{A} = \mathbf{A}$.

A generalized inverse is readily obtained if **A** is expressed in canonical form. Let the columns of **A** be spanned by a set of normalized eigenvectors $\mathbf{p}_1, \mathbf{p}_2, \ldots, \mathbf{p}_v$ with corresponding eigenvalues $\lambda_1, \lambda_2, \ldots, \lambda_v$; $\mathbf{p}_i'\mathbf{p}_i = 1$ and $\mathbf{p}_i'\mathbf{p}_j = 0$ $(i \neq j)$. Then, in canonical form

$$\mathbf{A} = \sum_{i=1}^{v} \lambda_i \mathbf{p}_i \mathbf{p}_i'. \qquad (1.16)$$

It can be verified that

$$\boldsymbol{\Omega} = \sum \lambda_i^{-1} \mathbf{p}_i \mathbf{p}_i' \qquad (1.17)$$

is a generalized inverse of **A**, where the summation is over all the non-zero eigenvalues.

Since $\hat{\boldsymbol{\tau}}'\mathbf{1} = 0$, it is suggested that $\hat{\mu} = \bar{y}$ is added to each element in $\hat{\boldsymbol{\tau}}$ so that the treatment estimates from the intra-block analysis are then directly comparable with the unadjusted treatment means $\mathbf{r}^{-\delta}\mathbf{T}$. Hence, the estimates are

$$\hat{\boldsymbol{\tau}}^* = \hat{\boldsymbol{\tau}} + \bar{y}\mathbf{1} \qquad (1.18)$$

where $\hat{\boldsymbol{\tau}}$ is obtained from (1.15) with $\boldsymbol{\Omega}$ given by (1.17).

An alternative generalized inverse to that given in (1.17), and one that is often easier to obtain, is given in Section 1.6 for designs for which rank $(\mathbf{A}) = v - 1$.

Example 1.1 (continued)

$$\mathbf{N}\mathbf{k}^{-\delta}\mathbf{N}' = \frac{1}{6}\begin{pmatrix} 8 & 5 & 5 \\ 5 & 8 & 5 \\ 5 & 5 & 8 \end{pmatrix}, \qquad \mathbf{N}\mathbf{k}^{-\delta}\mathbf{B} = \frac{1}{6}\begin{pmatrix} 199 \\ 175 \\ 190 \end{pmatrix}$$

Hence

$$\mathbf{A} = \mathbf{r}^{\delta} - \mathbf{N}\mathbf{k}^{-\delta}\mathbf{N}' = \frac{5}{6}\begin{pmatrix} 2 & -1 & -1 \\ -1 & 2 & -1 \\ -1 & -1 & 2 \end{pmatrix}$$

$$\mathbf{q} = \mathbf{T} - \mathbf{N}\mathbf{k}^{-\delta}\mathbf{B} = \frac{1}{6}\begin{pmatrix} 19 \\ -23 \\ 4 \end{pmatrix}$$

Normalized eigenvectors of **A** are given by

$$\mathbf{p}_1 = 2^{-1/2}\begin{pmatrix} 1 \\ 0 \\ -1 \end{pmatrix}, \quad \mathbf{p}_2 = 6^{-1/2}\begin{pmatrix} 1 \\ -2 \\ 1 \end{pmatrix}, \quad \mathbf{p}_3 = 3^{-1/2}\begin{pmatrix} 1 \\ 1 \\ 1 \end{pmatrix}$$

with eigenvalues $\lambda_1 = 15/6$, $\lambda_2 = 15/6$, $\lambda_3 = 0$. Therefore

$$\mathbf{\Omega} = \frac{6}{15}\left[\frac{1}{2}\begin{pmatrix} 1 \\ 0 \\ -1 \end{pmatrix}(1 \;\; 0 \;\; -1) + \frac{1}{6}\begin{pmatrix} 1 \\ -2 \\ 1 \end{pmatrix}(1 \;\; -2 \;\; 1)\right]$$

$$= \frac{2}{15}\begin{pmatrix} 2 & -1 & -1 \\ -1 & 2 & -1 \\ -1 & -1 & 2 \end{pmatrix}$$

and

$$\hat{\boldsymbol{\tau}} = \mathbf{\Omega}\mathbf{q} = \frac{1}{45}\begin{pmatrix} 57 \\ -69 \\ 12 \end{pmatrix}.$$

Finally,

$$\hat{\mu} = \bar{y} = \frac{94}{9}$$

and

$$\hat{\boldsymbol{\beta}} = \mathbf{k}^{-\delta}(\mathbf{B} - \mathbf{N}'\boldsymbol{\tau}) - \mathbf{1}\bar{y} = \frac{1}{90}\begin{pmatrix} -364 \\ 32 \\ 290 \\ -247 \end{pmatrix}$$

1.4 Estimability

Properties of the parameter estimators obtained from a solution of the normal equations will now be examined. First, the expected value and variance–covariance matrix of the adjusted treatment vector

$\mathbf{q}$ will be obtained. Let the model (1.3) be premultiplied successively by $\mathbf{X}'$ and $\mathbf{Nk}^{-\delta}\mathbf{Z}'$ to give

$$\mathbf{X}'\mathbf{y} = \mathbf{r}\mu + \mathbf{r}^{\delta}\boldsymbol{\tau} + \mathbf{N}\boldsymbol{\beta} + \mathbf{X}'\boldsymbol{\varepsilon} \tag{1.19}$$

$$\mathbf{Nk}^{-\delta}\mathbf{Z}'\mathbf{y} = \mathbf{r}\mu + \mathbf{Nk}^{-\delta}\mathbf{N}'\boldsymbol{\tau} + \mathbf{N}\boldsymbol{\beta} + \mathbf{Nk}^{-\delta}\mathbf{Z}'\boldsymbol{\varepsilon} \tag{1.20}$$

where again use has been made of (1.4)–(1.6). Subtracting (1.20) from (1.19) gives

$$\mathbf{q} = \mathbf{A}\boldsymbol{\tau} + (\mathbf{X}' - \mathbf{Nk}^{-\delta}\mathbf{Z}')\boldsymbol{\varepsilon} \tag{1.21}$$

Since $E(\boldsymbol{\varepsilon}) = \mathbf{0}$ and $V(\boldsymbol{\varepsilon}) = \sigma^2\mathbf{I}$ it follows that

$$E(\mathbf{q}) = \mathbf{A}\boldsymbol{\tau}$$

and

$$\begin{aligned} V(\mathbf{q}) &= (\mathbf{X}' - \mathbf{Nk}^{-\delta}\mathbf{Z}')(\mathbf{X}' - \mathbf{Nk}^{-\delta}\mathbf{Z}')'\sigma^2 \\ &= (\mathbf{r}^{\delta} - \mathbf{Nk}^{-\delta}\mathbf{N}')\sigma^2 = \mathbf{A}\sigma^2 \end{aligned}$$

so that, using (1.15),

$$\begin{aligned} E(\hat{\boldsymbol{\tau}}) &= \boldsymbol{\Omega}\mathbf{A}\boldsymbol{\tau} \\ V(\hat{\boldsymbol{\tau}}) &= \boldsymbol{\Omega}\mathbf{A}\boldsymbol{\Omega}\sigma^2 \end{aligned} \tag{1.22}$$

Hence $\hat{\boldsymbol{\tau}}$ is an unbiased estimator of a function of the treatment parameters. Further, this function depends on the non-unique generalized inverse matrix $\boldsymbol{\Omega}$. Although the treatment parameters themselves are not uniquely estimated and do not have unbiased estimators, certain linear functions of these treatment parameters are *estimable.* In general, a linear function $\mathbf{c}'\boldsymbol{\tau}$ is said to be estimable if there exists some linear combination of the responses $\mathbf{b}'\mathbf{y}$ such that $E(\mathbf{b}'\mathbf{y}) = \mathbf{c}'\boldsymbol{\tau}$. Since $\hat{\boldsymbol{\tau}}$ is a linear function of the responses it follows from (1.22) that $\mathbf{c}'\boldsymbol{\tau}$ is estimable if the coefficient vector $\mathbf{c}$ is chosen to satisfy $\mathbf{c}' = \mathbf{c}'\boldsymbol{\Omega}\mathbf{A}$. This *estimability condition* is, in fact, both necessary and sufficient for $\mathbf{c}'\boldsymbol{\tau}$ to be estimable; for a fuller discussion of estimability; see Searle (1971, p. 180). The unbiased estimator of the estimable function $\mathbf{c}'\boldsymbol{\tau}$ is $\mathbf{c}'\hat{\boldsymbol{\tau}}$ and, further, this estimator is invariant to whatever solution of the reduced normal equations (1.11) is used for $\hat{\boldsymbol{\tau}}$. Finally, from (1.22) it follows that the variance of this estimator is given by

$$\operatorname{var}(\mathbf{c}'\hat{\boldsymbol{\tau}}) = \mathbf{c}'\boldsymbol{\Omega}\mathbf{c}\sigma^2 \tag{1.23}$$

As an example of an estimable function, consider again the canonical forms of $\mathbf{A}$ and $\boldsymbol{\Omega}$ given in (1.16) and (1.17) respectively.

Let $\mathbf{p}_1$ be an eigenvector with non-zero eigenvalue λ_1. It follows immediately that $\mathbf{p}_1' = \mathbf{p}_1'\mathbf{\Omega A}$ so that $\mathbf{p}_1'\boldsymbol{\tau}$ is estimable with unbiased estimator $\lambda_1^{-1}\mathbf{p}_1'\mathbf{q}$, whose variance is $\lambda_1^{-1}\sigma^2$.

1.5 Analysis of variance

The further assumption that the error terms $\boldsymbol{\varepsilon}$ in (1.3) are normally distributed is required when carrying out significance tests or setting confidence intervals. It will be assumed, therefore, that the error vector $\boldsymbol{\varepsilon}$ follows a multivariate normal distribution with mean vector $\mathbf{0}$ and variance–covariance matrix $\sigma^2\mathbf{I}$.

The total sum of squares $\mathbf{y}'\mathbf{y}$, for the general linear model $\mathbf{y} = \mathbf{W}\boldsymbol{\alpha} + \boldsymbol{\varepsilon}$, can be partitioned into two components as follows:

$$\mathbf{y}'\mathbf{y} = \hat{\boldsymbol{\alpha}}'\mathbf{W}'\mathbf{y} + (\mathbf{y} - \mathbf{W}\hat{\boldsymbol{\alpha}})'(\mathbf{y} - \mathbf{W}\hat{\boldsymbol{\alpha}}) \tag{1.24}$$

where $\hat{\boldsymbol{\alpha}}$ is any solution to the normal equations $\mathbf{W}'\mathbf{W}\hat{\boldsymbol{\alpha}} = \mathbf{W}'\mathbf{y}$. Since $\mathbf{y} - \mathbf{W}\hat{\boldsymbol{\alpha}}$ is the vector of residuals, the second component in (1.24) is the residual sum of squares. The first component, therefore, represents that part of the total variation accounted for by the model. That is, the sums of squares due to fitting all the parameters in the model $\mathbf{y} = \mathbf{W}\boldsymbol{\alpha} + \boldsymbol{\varepsilon}$ is given by $\hat{\boldsymbol{\alpha}}'\mathbf{W}'\mathbf{y} = S(\boldsymbol{\alpha})$. Note that this sum of squares is the sum of cross-products of the least squares estimator $\hat{\boldsymbol{\alpha}}$ and the terms on the right-hand side of the normal equations $\mathbf{W}'\mathbf{y}$.

Hence, the sum of squares due to fitting the parameters in the intra-block model (1.3) is given by

$$S(\mu, \boldsymbol{\tau}, \boldsymbol{\beta}) = \hat{\mu}G + \hat{\boldsymbol{\tau}}'\mathbf{T} + \hat{\boldsymbol{\beta}}'\mathbf{B}$$

which, using (1.13) and (1.14), becomes

$$S(\mu, \boldsymbol{\tau}, \boldsymbol{\beta}) = \hat{\boldsymbol{\tau}}'\mathbf{q} + \mathbf{B}'\mathbf{k}^{-\delta}\mathbf{B} \tag{1.25}$$

Suppose now that the hypothesis of no significant differences between treatment parameters is to be tested. That is, the null hypothesis of interest is

$$H_0: \tau_1 = \tau_2 = \cdots = \tau_v$$

Under the assumption that H_0 is true, model (1.3) reduces to the one-way model given by

$$\mathbf{y} = \mathbf{1}\mu + \mathbf{Z}'\boldsymbol{\beta} + \boldsymbol{\varepsilon} \tag{1.26}$$

The normal equations for this model are

$$n\hat{\mu} + \mathbf{k}'\hat{\boldsymbol{\beta}} = G$$
$$\mathbf{k}\hat{\mu} + \mathbf{k}^{\delta}\hat{\boldsymbol{\beta}} = \mathbf{B}$$

so that with $\hat{\mu} = \bar{y}$,

$$\hat{\boldsymbol{\beta}} = \mathbf{k}^{-\delta}\mathbf{B} - \mathbf{1}\bar{y}$$

Thus

$$S(\mu, \boldsymbol{\beta}) = \hat{\mu}G + \hat{\boldsymbol{\beta}}'\mathbf{B} = \mathbf{B}'\mathbf{k}^{-\delta}\mathbf{B} \tag{1.27}$$

The difference between the two sums of squares (1.25) and (1.27) is the sum of squares due to testing the hypothesis H_0. Since it is the sum of squares due to treatment effects after eliminating the overall mean parameter and the block parameters, it will be denoted by $S(\boldsymbol{\tau}/\mu, \boldsymbol{\beta})$. Hence

$$S(\boldsymbol{\tau}/\mu, \boldsymbol{\beta}) = S(\mu, \boldsymbol{\tau}, \boldsymbol{\beta}) - S(\mu, \boldsymbol{\beta}) = \hat{\boldsymbol{\tau}}'\mathbf{q} \tag{1.28}$$

The sum of squares $S(\mu, \boldsymbol{\beta})$ in (1.28) can itself be partitioned further. Putting all block parameters equal in (1.26) leads to the reduced model $\mathbf{y} = \mathbf{1}\mu + \boldsymbol{\varepsilon}$. From the normal equations for this model it follows that $\hat{\mu} = \bar{y}$ and $S(\mu) = G^2/n$. Hence

$$S(\boldsymbol{\beta}/\mu) = S(\mu, \boldsymbol{\beta}) - S(\mu) = \mathbf{B}'\mathbf{k}^{-\delta}\mathbf{B} - G^2/n \tag{1.29}$$

$S(\boldsymbol{\beta}/\mu)$ is the sum of squares due to blocks after eliminating the mean. Since it does not refer at all to treatments it represents between-block variation; it can be seen in (1.29) to be a function of the block totals $\mathbf{B}$. $S(\boldsymbol{\tau}/\mu, \boldsymbol{\beta})$, on the other hand, represents variation within blocks since block effects have been eliminated.

The above results are summarized in the *analysis of variance* given in Table 1.3. It should be noted that all treatment differences must be estimable in the design if H_0 is to be tested; this point will be discussed in Section 1.6. The residual mean square s^2 provides an unbiased estimator of σ^2 and is used as the estimate of experimental error in any significance test and in setting confidence limits. The hypothesis H_0 is then tested by comparing the variance ratio

$$F = \frac{\hat{\boldsymbol{\tau}}'\mathbf{q}/(v-1)}{s^2}$$

with the appropriate percentage points of the F distribution with $v - 1$ and $n - b - v + 1$ degrees of freedom.

Table 1.3 *Intra-block analysis of variance.*

Source of variation	*Degrees of freedom (d.f.)*	*Sum of squares (s.s.)*	*Mean square (m.s.)*
Between blocks (ignoring treatments)	$b-1$	$S(\boldsymbol{\beta}/\mu)=\mathbf{B}'k^{-\delta}\mathbf{B} - G^2/n$	
Between treatments (eliminating blocks)	$v-1$	$S(\boldsymbol{\tau}/\mu,\boldsymbol{\beta})=\hat{\boldsymbol{\tau}}'\mathbf{q}$	$\hat{\boldsymbol{\tau}}'\mathbf{q}/(v-1)$
Residual	$n-b-v+1$	$\mathbf{y}'\mathbf{y}-S(\mu,\boldsymbol{\tau},\boldsymbol{\beta})$	s^2
Total (corrected for the mean)	$n-1$	$\mathbf{y}'\mathbf{y}-G^2/n$	

To produce the analysis of variance given in Table 1.3 a sequence of models was fitted, starting with a mean term then adding in turn the block parameters and the treatment parameters. This led to the following decomposition of the sum of squares $S(\mu,\boldsymbol{\tau},\boldsymbol{\beta})$ due to fitting the intra-block model (1.3):

$$S(\mu,\boldsymbol{\tau},\boldsymbol{\beta})=S(\mu)+S(\boldsymbol{\beta}/\mu)+S(\boldsymbol{\tau}/\mu,\boldsymbol{\beta}) \tag{1.30}$$

This particular model fitting sequence was followed as interest is centred on the treatments in a blocking experiment and, thus, it is necessary to estimate and test the treatment parameters in the absence of, or after adjusting for, the block effects.

However, adding treatment parameters to the model before the block parameters would lead to another analysis of variance and the following decomposition:

$$S(\mu,\boldsymbol{\tau},\boldsymbol{\beta})=S(\mu)+S(\boldsymbol{\tau}/\mu)+S(\boldsymbol{\beta}/\mu,\boldsymbol{\tau}) \tag{1.31}$$

Analogous to (1.29),

$$S(\boldsymbol{\tau}/\mu)=\mathbf{T}'\mathbf{r}^{-\delta}\mathbf{T}-G^2/n \tag{1.32}$$

and represents the unadjusted treatment sum of squares (between treatments, ignoring blocks). By subtraction from (1.30) and (1.31) the adjusted block sum of squares (between blocks, eliminating treatments) is given by

$$S(\boldsymbol{\beta}/\mu,\boldsymbol{\tau})=S(\boldsymbol{\beta}/\mu)+S(\boldsymbol{\tau}/\mu,\boldsymbol{\beta})-S(\boldsymbol{\tau}/\mu) \tag{1.33}$$

Alternatively, following through similar algebra to that used to obtain (1.28), it can be shown that

$$S(\boldsymbol{\beta}/\mu, \tau) = \hat{\boldsymbol{\beta}}'\mathbf{s} \tag{1.34}$$

where $\mathbf{s}$ is the vector of adjusted block totals given by

$$\mathbf{s} = \mathbf{B} - \mathbf{N}'\mathbf{r}^{-\delta}\mathbf{T}$$

and $\hat{\boldsymbol{\beta}}$ is a solution to the set of equations

$$(\mathbf{k}^{\delta} - \mathbf{N}'\mathbf{r}^{-\delta}\mathbf{N})\hat{\boldsymbol{\beta}} = \mathbf{s}.$$

In a blocking experiment the second decomposition given in (1.31) is of little importance in itself since it is rarely of interest to examine block effects in detail. However, the sums of squares given in (1.32), (1.33) and (1.34) will be needed in the discussions of orthogonality in Section 1.7 and of the recovery of inter-block information in Chapter 8; it is convenient to define them here.

Example 1.1 (continued)
Using the data and calculations of Sections 1.2 and 1.3,

$$\mathbf{y}'\mathbf{y} = 8^2 + 14^2 + 9^2 + 5^2 + 13^2 + 16^2 + 12^2 + 9^2 + 8^2 = 1080$$

$$S(\mu) = G^2/n = 94^2/9 = 981.78$$

$$S(\boldsymbol{\beta}/\mu) = \frac{22^2}{2} + \frac{14^2}{2} + \frac{41^2}{3} + \frac{17^2}{2} - \frac{G^2}{n} = 63.06$$

$$S(\tau/\mu, \boldsymbol{\beta}) = \frac{1}{45 \times 6}(57 \quad -69 \quad 12)\begin{pmatrix} 19 \\ -23 \\ 4 \end{pmatrix} = 10.07.$$

Table 1.4 *Analysis of variance table.*

	d.f.	*s.s.*	*m.s.*
Between blocks (ignoring treatments)	3	63.06	21.02
Between treatments (eliminating blocks)	2	10.07	5.04
Residual	3	25.09	8.36
Total (corrected)	8	98.22	

The intra-block analysis of variance is shown in Table 1.4. The variance-ratio statistic to test $H_0\colon \tau_1 = \tau_2 = \tau_3$ is

$$F = \frac{5.04}{8.36} = 0.6$$

based on 2 and 3 degrees of freedom. As this is a small hypothetical example with very few degrees of freedom for error, no conclusions should be drawn from this analysis. The above arithmetic is purely to illustrate the calculations involved.

Finally, the unadjusted treatment sum of squares and the adjusted block sum of squares are given by

$$S(\boldsymbol{\tau}/\mu) = \frac{30^2 + 33^2 + 31^2}{3} - \frac{G^2}{n} = 1.56$$

$$S(\boldsymbol{\beta}/\mu, \boldsymbol{\tau}) = 63.06 + 10.07 - 1.56 = 71.57$$

Hypotheses other than H_0 involving the treatment parameters may be of interest. It may be desired to test the hypothesis that some estimable function $\mathbf{c}'\boldsymbol{\tau}$ is equal to some given quantity. More generally, the hypothesis may involve a number of such estimable functions. Suppose that $\mathbf{C}$ is a matrix such that each of the rows of $\mathbf{C}\boldsymbol{\tau}$ is estimable and linearly independent of the other rows. Then the sum of squares due to testing the hypothesis $H_0(\mathbf{C})\colon \mathbf{C}\boldsymbol{\tau} = \mathbf{m}$, where $\mathbf{m}$ is a vector of known quantities, is given by

$$S[H_0(\mathbf{C})] = (\mathbf{C}\hat{\boldsymbol{\tau}} - \mathbf{m})'(\mathbf{C}\boldsymbol{\Omega}\mathbf{C}')^{-1}(\mathbf{C}\hat{\boldsymbol{\tau}} - \mathbf{m}) \qquad (1.35)$$

based on ν degrees of freedom, where $\nu = \text{rank}(\mathbf{C})$; see Searle (1971, p. 190). In Chapters 6 and 7 on factorial experiments it will be more convenient to work with $\mathbf{C}$ matrices which do not have full row rank. In such cases, the sum of squares is again given by (1.35) with the inverse of $\mathbf{C}\boldsymbol{\Omega}\mathbf{C}'$ now replaced by a generalized inverse of $\mathbf{C}\boldsymbol{\Omega}\mathbf{C}'$; see John and Smith (1974).

1.6 Connectedness

The $v \times v$ information matrix $\mathbf{A}$ given in (1.12) has rank less than or equal to $v - 1$. The equality will hold if and only if the design is *connected*. In the discussion which follows attention is initially restricted to *binary* designs which have equal replication and equal block sizes, i.e. $r_i = r$ and $k_j = k$ for all i and j. A binary design is one in which each element of the incidence matrix $\mathbf{N}$ is either 0 or 1. For such designs the information matrix simplifies to

$$\mathbf{A} = r\mathbf{I} - (1/k)\mathbf{NN}' \tag{1.36}$$

where the treatment *concurrence matrix* $\mathbf{NN}'$ has diagonal elements equal to r and off-diagonal elements equal to the number of times pairs of treatments occur together in a block. Hence, if $\mathbf{A} = ((a_{ij}))$ then

$$a_{ij} = \begin{cases} r(k-1)/k, & i = j \\ -\lambda_{ij}/k, & i \neq j \end{cases}$$

where λ_{ij} gives the number of blocks containing both the ith and jth treatments.

Now suppose that $\text{rank}(\mathbf{A}) < v - 1$. Then there exists some vector $\mathbf{x}$, other than the unit vector $\mathbf{1}$ and the zero vector $\mathbf{0}$, such that $\mathbf{Ax} = \mathbf{0}$. By adding or subtracting multiples of $\mathbf{1}$, it is possible to assume that one of the elements in $\mathbf{x}$ is equal to zero while all other elements are greater than or equal to zero. Without loss of generality, let $x_1 = 0$ and $x_j \geqslant 0$ for $j > 1$. The first row of $\mathbf{Ax} = \mathbf{0}$ then gives

$$\sum_{j=2}^{v} a_{1j}x_j = 0 \qquad \text{or} \qquad \sum_{j=2}^{v} \lambda_{1j}x_j = 0$$

Now both $x_j \geqslant 0$ and $\lambda_{1j} \geqslant 0$, for all $j > 1$. If all $x_j > 0$ then $\lambda_{1j} = 0$ for all $j > 1$ which means that the first treatment is in a block of its own, hence $k = 1$. Without loss of generality, therefore, let $x_2 = x_3 = \cdots = x_{s-1} = 0$ where $s - 1 < v$, since not all x_j can be zero as $\mathbf{x}$ is not a multiple of the unit vector. Then, it follows that $\lambda_{1s} = \cdots = \lambda_{1v} = 0$. Now the lth row of $\mathbf{Ax} = \mathbf{0}$ gives, for $l = 2, 3, \ldots, s-1$,

$$\sum_{j=s}^{v} \lambda_{lj}x_j = 0 \qquad \text{which implies} \qquad \lambda_{lj} = 0 \quad \text{for} \quad j \geqslant s$$

This means that the 1st, 2nd, ..., $(s-1)$th treatments do not occur in any block with the remaining treatments. Such a design is said to be *disconnected*. More generally, a design is disconnected if the treatments can be split into groups such that no treatment from one group occurs in any block with any treatment from a different group. A design which is not disconnected is said to be *connected*. It follows that if $\text{rank}(\mathbf{A}) < v - 1$ then the design is disconnected. If the design is connected then $\text{rank}(\mathbf{A}) \geqslant v - 1$, which implies that $\text{rank}(\mathbf{A}) = v - 1$.*

*This proof due to Dr R.G. Jones, Australian National University, Canberra.

Example 1.2
Consider the following two designs for 6 treatments labelled $0, 1, \ldots, 5$ in 6 blocks of size 2 ($k = 2$) with each treatment replicated 2 times ($r = 2$).

Design 1 (0 2) (1 3) (2 4) (3 5) (4 0) (5 1)

Design 2 (0 1) (1 2) (2 3) (3 4) (4 5) (5 0)

Design 1 is disconnected with two groups of disconnected treatments 0, 2, 4 and 1, 3, 5. Design 2 is connected.

Further insight into the concept of connectedness can be obtained if designs are represented graphically. Let the points (vertices) of the graph represent treatments and let a line (edge) join two points if the corresponding treatments occur together in a block. Such a graph is called the *treatment concurrence graph* of the design since the number of lines joining any two points is given by the corresponding elements of the treatment concurrence matrix $\mathbf{NN}'$. The graphs for the designs in Example 1.2 are shown in Fig. 1.1. In a connected design it is possible to find a path from any point in the treatment concurrence graph to any other point; in a disconnected design this is not possible. In the graph for Design 1 no path joins any even-numbered point to any odd-numbered point; the design is thus disconnected. For Design 2, on the other hand, it is possible to traverse the lines of its graph so as to be able to move from any point to any other point; the design is thus connected.

Although the above proof on the rank of the matrix $\mathbf{A}$ refers to

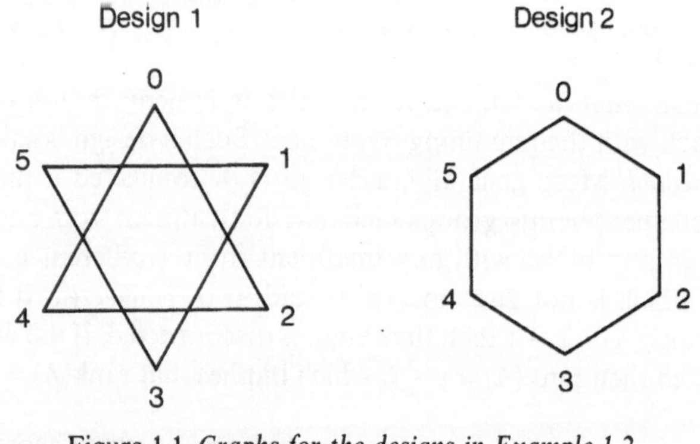

Figure 1.1 *Graphs for the designs in Example 1.2.*

binary, equal block size and equal replicate designs, the concept of connectedness applies equally well for non-binary designs and for designs with unequal block sizes and unequal replication. That is, *any* block design is disconnected if the blocks can be split into groups in such a way that the treatments in any one group of blocks are distinct from the treatments in the other groups. A design is connected if it is not disconnected.

Connectedness is important because in connected designs every treatment contrast is estimable from comparisons within blocks. The linear function $\mathbf{c}'\boldsymbol{\tau}$ is a *contrast* in the treatment parameters if the elements of the coefficient vector $\mathbf{c}$ sum to zero, i.e. $\mathbf{c}'\mathbf{1} = 0$. It is necessary to show, therefore, that all such contrast vectors $\mathbf{c}$ satisfy the estimability condition $\mathbf{c}' = \mathbf{c}'\boldsymbol{\Omega}\mathbf{A}$ given in Section 1.4.

A generalized inverse $\boldsymbol{\Omega}$ of the information matrix $\mathbf{A}$ has been given in (1.17). The required result can be readily proven using this inverse since, for a connected design, there will be $v-1$ orthogonal eigenvectors with non-zero eigenvalues spanning the $(v-1)$-dimensional contrast space. However, an alternative proof is now given as it provides a wider class of generalized inverses of $\mathbf{A}$ which will be found useful in later chapters.

For a connected design, let $\mathbf{p}_v = v^{-1/2}\mathbf{1}$ be the normalized eigenvector with zero eigenvalue. Hence the matrix $\mathbf{A} + a\mathbf{J}$ is nonsingular for all $a \neq 0$ since, using (1.16), $\mathbf{A} + a\mathbf{J} = \sum \lambda_i^* \mathbf{p}_i\mathbf{p}_i'$ so that $(\mathbf{A} + a\mathbf{J})^{-1} = \sum \lambda_i^{*-1}\mathbf{p}_i\mathbf{p}_i'$, where both summations are over $i = 1, 2, \ldots, v$ and where $\lambda_i^* = \lambda_i$ for $i = 1, 2, \ldots, v-1$, and $\lambda_v^* = av$. It is then readily verified that

$$\boldsymbol{\Omega} = (\mathbf{A} + a\mathbf{J})^{-1}, \qquad a \neq 0 \tag{1.37}$$

is a generalized inverse of $\mathbf{A}$. It also follows that

$$\boldsymbol{\Omega}\mathbf{A} = \sum_{i=1}^{v} \mathbf{p}_i\mathbf{p}_i' - \mathbf{p}_v\mathbf{p}_v' = \mathbf{I} - \mathbf{K} \tag{1.38}$$

where $\mathbf{K}$ is the $v \times v$ matrix with every element equal to v^{-1}. Hence $\mathbf{c}' = \mathbf{c}'\boldsymbol{\Omega}\mathbf{A}$ if and only if $\mathbf{c}'\mathbf{1} = 0$. That is, in a connected design, all contrasts in the treatment parameters are estimable.

The generalized inverse $\boldsymbol{\Omega}$ of (1.37) was given by Shah (1959). More generally, John (1965) showed that $\boldsymbol{\Omega} = (\mathbf{A} + a\mathbf{h}\mathbf{h}')^{-1}$ where $a \neq 0$ and $\mathbf{h}$ is any vector such that the columns of $\mathbf{A}$ together with $\mathbf{h}$ span a v-dimensional space. These results, with an appropriate choice of the constant a, will often provide a simple method of obtaining a solution to the reduced normal equations (1.11) for many different types of designs.

Note that if $\boldsymbol{\Omega}$ is obtained from (1.37) with any $a \neq 0$ or from (1.17), the $a = 0$ case, the same solution will be obtained to the normal equations. This is because

$$\hat{\boldsymbol{\tau}} = \boldsymbol{\Omega}\mathbf{q} = \sum_{i=1}^{v} \lambda_i^{*-1}\mathbf{p}_i\mathbf{p}_i'\mathbf{q} = \sum_{i=1}^{v-1} \lambda_i^{-1}\mathbf{p}_i\mathbf{p}_i'\mathbf{q}$$

since $\mathbf{1}'\mathbf{q} = 0$. This solution is independent of the constant a. With other generalized inverses, however, different solutions may be obtained although estimates of treatment contrasts will not differ since they are estimable.

Example 1.1 (continued)
From Section 1.3, the information matrix $\mathbf{A}$ and the vector of adjusted treatment totals $\mathbf{q}$ are

$$\mathbf{A} = \frac{5}{6}\begin{pmatrix} 2 & -1 & -1 \\ -1 & 2 & -1 \\ -1 & -1 & 2 \end{pmatrix}, \quad \mathbf{q} = \frac{1}{6}\begin{pmatrix} 19 \\ -23 \\ 4 \end{pmatrix}$$

A generalized inverse of $\mathbf{A}$, using (1.37) with $a = 5/6$, is given by $\boldsymbol{\Omega} = (6/15)\mathbf{I}$. Hence

$$\hat{\boldsymbol{\tau}} = \boldsymbol{\Omega}\mathbf{q} = \frac{1}{15}\begin{pmatrix} 19 \\ -23 \\ 4 \end{pmatrix}$$

as before. To illustrate the final point made above, consider the following generalized inverse of $\mathbf{A}$ which does not belong to the class given by (1.37), namely

$$\boldsymbol{\Omega} = \frac{6}{15}\begin{pmatrix} 2 & 1 & 0 \\ 1 & 2 & 0 \\ 0 & 0 & 0 \end{pmatrix}$$

This leads to the different solution

$$\hat{\boldsymbol{\tau}} = \frac{1}{15}\begin{pmatrix} 15 \\ -27 \\ 0 \end{pmatrix}$$

Note, however, that $\hat{\tau}_1 - \hat{\tau}_2 = 2.80$ and $\hat{\tau}_1 - \hat{\tau}_3 = 1.00$ for both of the above solutions. This will be the case for any contrasts in these estimates.

1.7 Orthogonality

In the linear model $\mathbf{y} = \mathbf{W}_1\boldsymbol{\alpha}_1 + \mathbf{W}_2\boldsymbol{\alpha}_2 + \boldsymbol{\varepsilon}$ the parameter vectors $\boldsymbol{\alpha}_1$ and $\boldsymbol{\alpha}_2$ are said to be *orthogonal* to each other if their least squares estimators are independent, in the sense that the estimator of $\boldsymbol{\alpha}_i$ in this model is the same as that obtained from the model $\mathbf{y} = \mathbf{W}_i\boldsymbol{\alpha}_i + \boldsymbol{\varepsilon}$ $(i = 1, 2)$. That is, the presence or absence of one of the parameter vectors in the model does not affect the estimator of the other vector. The analysis of such orthogonal models is consequently simplified. From a consideration of the normal equations, it is clear that $\boldsymbol{\alpha}_1$ and $\boldsymbol{\alpha}_2$ will be orthogonal if and only if $\mathbf{W}_1'\mathbf{W}_2 = \mathbf{0}$. The requirement for orthogonality in block designs will now be examined.

The block and treatment parameter vectors will be orthogonal to the mean parameter μ in model (1.3) if $\mathbf{X}$ and $\mathbf{Z}$ are replaced by $\mathbf{X}^* = \mathbf{X} - (1/n)\mathbf{1}\mathbf{r}'$ and $\mathbf{Z}^* = \mathbf{Z} - (1/n)\mathbf{1}\mathbf{k}'$ respectively. This ensures that $\mathbf{X}^{*\prime}\mathbf{1} = \mathbf{0}$ and $\mathbf{Z}^{*\prime}\mathbf{1} = \mathbf{0}$. For orthogonality between blocks and treatments it is then necessary that $\mathbf{X}^{*\prime}\mathbf{Z}^* = \mathbf{0}$. Simplifying this expression, using (1.4)–(1.6), the condition for orthogonality in a block design is seen to reduce to

$$\mathbf{N} = (1/n)\mathbf{r}\mathbf{k}' \tag{1.39}$$

With incidence matrix $\mathbf{N}$ given by (1.39), the information matrix $\mathbf{A}$ and vector of adjusted treatment totals $\mathbf{q}$, given in (1.12) and (1.13) respectively, become

$$\mathbf{A} = \mathbf{r}^{\delta} - (1/n)\mathbf{r}\mathbf{r}', \tag{1.40}$$

$$\mathbf{q} = \mathbf{T} - \bar{y}\mathbf{r}. \tag{1.41}$$

It is easily verified, by showing that $\mathbf{A}\boldsymbol{\Omega}\mathbf{A} = \mathbf{A}$, that a generalized inverse of $\mathbf{A}$ is given by

$$\boldsymbol{\Omega} = \mathbf{r}^{-\delta} \tag{1.42}$$

Hence

$$\hat{\boldsymbol{\tau}} = \boldsymbol{\Omega}\mathbf{q} = \mathbf{r}^{-\delta}\mathbf{T} - \bar{y}\mathbf{1} \tag{1.43}$$

Contrasts in the treatment parameters are estimable with estimates given by contrasts in the unadjusted treatment means. That is, $\mathbf{c}'\boldsymbol{\tau}$ is estimable if and only if $\mathbf{c}'\mathbf{1} = 0$, with estimator $\mathbf{c}'(\mathbf{r}^{-\delta}\mathbf{T})$.

The adjusted sum of squares due to treatment effects is

$$S(\boldsymbol{\tau}/\mu, \boldsymbol{\beta}) = \hat{\boldsymbol{\tau}}'\mathbf{q} = \mathbf{T}'\mathbf{r}^{-\delta}\mathbf{T} - G^2/n$$

which is the same as the unadjusted treatment sum of squares given in (1.32). Thus, for orthogonal block designs,

$$S(\boldsymbol{\tau}/\mu, \boldsymbol{\beta}) = S(\boldsymbol{\tau}/\mu) \tag{1.44}$$

In a similar way it can be shown that $S(\boldsymbol{\beta}/\mu, \boldsymbol{\tau}) = S(\boldsymbol{\beta}/\mu)$.

Orthogonality between treatments and blocks has resulted, therefore, in a considerable simplification in the analysis. Estimates of treatment parameters are based on treatment means, and no adjustment for block totals is necessary. It is also unnecessary to eliminate block effects in calculating the treatment sum of squares. As a consequence, all the information on treatment effects is contained in comparisons within blocks. Similarly, block parameters can be estimated and tested without the need to eliminate treatment effects.

The construction of orthogonal designs follows immediately from (1.39) since knowledge of the incidence matrix $\mathbf{N}$ gives the design directly; remember that the (ij)th element of $\mathbf{N}$ gives the number of times the ith treatment occurs in the jth block.

Example 1.3

An orthogonal design for $v=2$ treatments, labelled 0, 1, with $\mathbf{r} = (3,3)'$, $\mathbf{k} = (2,4)'$ and $n = \mathbf{r}'\mathbf{1} = \mathbf{k}'\mathbf{1} = 6$ has, from (1.39),

$$\mathbf{N} = \begin{pmatrix} 1 & 2 \\ 1 & 2 \end{pmatrix}$$

Thus, both treatments 0 and 1 occur once in the first block and twice in the second block. The two blocks of this design are, therefore, (0 1) and (0 0 1 1).

Of particular importance are the orthogonal designs for equal block sizes and equal treatment replication. It follows from (1.39) that $\mathbf{N}$ must equal $p\mathbf{J}$, for some positive integer p. For orthogonality, therefore, each treatment occurs the same number of times, p, in each block. Such a design is called a *complete* block design, and has $r = pb$ and $k = pv$ where b is the number of blocks and v the number of treatments. The design with $p = 1$ is called a *randomized block design.*

CHAPTER 2

Efficiency factors

2.1 Introduction

In an orthogonal design all the information on the treatments is obtained from the intra-block analysis. That is, contrasts among the treatment parameters are estimated entirely from comparisons made within blocks. In a non-orthogonal design some of the information is obtained from these within-block comparisons, but the remaining information has to be recovered, using an inter-block analysis, from comparison between blocks. Different designs of the same size can be assessed, therefore, on the basis of the amount of information available from both within and between blocks. For this purpose *efficiency factors*, obtained from the intra-block analysis, will be defined and used to provide criteria for comparing different non-orthogonal block designs.

Attention will be mainly restricted to designs in which the v treatments are set out in b blocks each of size k, such that each treatment is replicated r times. The efficiency factors of a block design are then obtained from a comparison of the variances of estimated treatment contrasts in the design with those in an orthogonal design using the same number of experimental units, under the assumption that the error variance σ^2 is the same for both designs. This is equivalent to making comparisons against the design that would have been obtained if no blocking had been used. The case of unequal replication will be discussed in Section 2.7.

The concept of an efficiency factor should not be confused with the measure of the gain (or loss) resulting from the use of a block design or of a design of a given size, the so called *efficiency* of a design. The purpose of blocking is to reduce the error variance σ^2 by exercising greater local control on the experiment; the gain in efficiency resulting from blocking will often be substantial. There will also frequently be a gain in efficiency resulting from using designs with small rather than large block sizes. However, designs of the same size will often differ in the precision with which treatment effects are estimated; it is the purpose of efficiency factors to provide convenient measures of these differences.

2.2 Pairwise efficiency factors

In a connected design all pairwise comparisons $\tau_i - \tau_j$ $(i \neq j)$ are estimable with estimators $\hat{\tau}_i - \hat{\tau}_j$, where $\hat{\tau}_i$ is the least squares estimator of τ_i obtained from (1.15). Further, from (1.23),

$$\mathrm{var}(\hat{\tau}_i - \hat{\tau}_j) = (\omega_{ii} + \omega_{jj} - 2\omega_{ij})\sigma^2 = v_{ij}\sigma^2 \quad (i \neq j),$$

where ω_{ij} is the (ij)th element of a generalized inverse $\mathbf{\Omega}$ of the information matrix $\mathbf{A}$ given in (1.12). For a complete block design this variance is, from (1.42), $2\sigma^2/r$ for all $i \neq j$. The efficiency factor E_{ij} for the pairwise comparison $\tau_i - \tau_j$ is then defined to be

$$E_{ij} = \frac{2\sigma^2/r}{v_{ij}\sigma^2} = \frac{2}{rv_{ij}} \tag{2.1}$$

The average variance of all estimate pairwise comparison is

$$\bar{v} = \frac{2\sigma^2}{v(v-1)} \sum_{i>j}\sum v_{ij}$$

so that an overall average efficiency factor can be defined by

$$E = \frac{2\sigma^2/r}{\bar{v}} = \frac{v(v-1)}{r\sum_{i>j}\sum v_{ij}} \tag{2.2}$$

Note that E is also the harmonic mean of the pairwise efficiency factors E_{ij} given in (2.1).

In an experiment it is often of interest to look for differences between pairs of treatments. Hence the pairwise efficiency factors E_{ij} are useful measures against which to judge the suitability of different designs. The distribution of the E_{ij} values, or their range or minimum (or maximum) value, give important information relating to the usefulness of any design. If all differences are of equal importance then the average efficiency factor E provides a convenient summary of the E_{ij}.

2.3 Canonical efficiency factors

The information matrix **A** has a complete set of orthogonal eigenvectors since it is a symmetric matrix. If there are multiplicities among the eigenvalues then the eigenvectors will not be uniquely determined, although a set of linearly independent eigenvectors can be found which will span the row or column space of **A**. Further, the eigenvectors of **A** can be normalized so that the sum of squares of the elements of any eigenvector is unity.

Let $\lambda_1, \lambda_2, \ldots, \lambda_v$ be the eigenvalues of **A**, not necessarily all distinct, with corresponding eigenvectors $\mathbf{p}_1, \mathbf{p}_2, \ldots, \mathbf{p}_v$ satisfying

$$\mathbf{p}_i'\mathbf{p}_j = \begin{cases} 1, & i = j \\ 0, & i \neq j \end{cases} \tag{2.3}$$

Then, as previously given in (1.16), **A** can be expressed in canonical form as

$$\mathbf{A} = \sum_{i=1}^{v} \lambda_i \mathbf{p}_i \mathbf{p}_i' \tag{2.4}$$

Since **A** is a singular matrix, at least one eigenvalue, λ_v say, will be zero with corresponding eigenvector $\mathbf{p}_v = (v^{-1/2})\mathbf{1}$. For a connected design all other $v - 1$ eigenvalues are non-zero, so that a generalized inverse of **A** is then given by

$$\mathbf{\Omega} = \sum_{i=1}^{v-1} \lambda_i^{-1} \mathbf{p}_i \mathbf{p}_i' \tag{2.5}$$

Attention will again be restricted to connected designs; for disconnected designs the main interest would in any case centre on identifying the non-estimable treatment contrasts rather than on any notion of efficiency factors.

Since $\mathbf{p}_j'\mathbf{p}_v = 0$ $(j \neq v)$, it follows that $\mathbf{p}_j'\boldsymbol{\tau}$ represents an estimable contrast in the treatment parameters. Its estimator is $\mathbf{p}_j'\hat{\boldsymbol{\tau}} = \lambda_j^{-1}\mathbf{p}_j'\mathbf{q}$, where $\mathbf{q}$ is the vector of adjusted treatment totals given in (1.13). Further, using (1.23),

$$\operatorname{var}(\mathbf{p}_j'\hat{\boldsymbol{\tau}}) = \mathbf{p}_j'\mathbf{\Omega}\mathbf{p}_j\sigma^2 = \sigma^2/\lambda_j$$

For a complete block design this variance is σ^2/r, using (1.42). Hence, the efficiency factor e_j of the contrast $\mathbf{p}_j'\boldsymbol{\tau}$ is

$$e_j = \lambda_j/r \qquad (j = 1, 2, \ldots, v - 1) \tag{2.6}$$

Note that the e_j in (2.6) are eigenvalues of the matrix

$$\mathbf{A}^* = (1/r)\mathbf{A} = \mathbf{I} - \mathbf{N}\mathbf{N}'/rk \tag{2.7}$$

The contrasts $\mathbf{p}_j'\boldsymbol{\tau}$ have been called the *basic contrasts* of a design by Pearce, Calinski and Marshall (1974). The non-zero eigenvalues of the matrix $\mathbf{A}^*$ in (2.7) are the efficiency factors of the basic contrasts, and have been called the *canonical efficiency factors* by James and Wilkinson (1971).

The adjusted treatment sum of squares (1.28) can now be written as

$$\hat{\boldsymbol{\tau}}'\mathbf{q} = \mathbf{q}'\boldsymbol{\Omega}\mathbf{q} = (1/r)\sum_{i=1}^{v-1} e_i^{-1}(\mathbf{p}_i'\mathbf{q})^2 \tag{2.8}$$

This means that the basic contrasts provide an orthogonal decomposition of the $v-1$ degrees of freedom for treatments into single degree of freedom components. If there are multiplicities among the canonical efficiency factors then this decomposition will not be unique. In particular, if all canonical efficiency factors are equal, as in a complete block design, then any set of $v-1$ linearly independent orthogonal contrasts will provide a decomposition of the treatment sum of squares.

Two important properties of canonical efficiency factors will now be considered. The first is that their harmonic mean is equal to the average efficiency factor E based on pairwise treatment differences given in (2.2). Let $\mathbf{d}_{jk}$ be the vector with jth element equal to 1, the kth element equal to -1 and the remaining $v-2$ elements equal to zero. Then

$$\operatorname{var}(\hat{\tau}_j - \hat{\tau}_k) = \mathbf{d}_{jk}'\boldsymbol{\Omega}\mathbf{d}_{jk}\sigma^2 = (\sigma^2/r)\sum_{i=1}^{v-1} e_i^{-1}(\mathbf{d}_{jk}'\mathbf{p}_i)^2$$

The average variance over all pairwise differences is, with $w = v(v-1)/2$,

$$\begin{aligned}\bar{v} &= (1/w)\sum_{j<k}\sum \operatorname{var}(\hat{\tau}_j - \hat{\tau}_k)\\ &= (\sigma^2/rw)\sum_i e_i^{-1}\Big[\sum_{j<k}\sum(\mathbf{d}_{jk}'\mathbf{p}_i)^2\Big]\\ &= (\sigma^2/rw)\sum_i e_i^{-1}[v\mathbf{p}_i'\mathbf{p}_i - (\mathbf{p}_i'\mathbf{1})^2]\\ &= (2\sigma^2/r)\left[\sum_{i=1}^{v-1} e_i^{-1}/(v-1)\right]\end{aligned}$$

The average efficiency factor E is then given by

$$E = \frac{v-1}{\sum_{i=1}^{v-1} e_i^{-1}} \tag{2.9}$$

namely the harmonic mean of the canonical efficiency factors.

The second property is that no treatment contrast can have an efficiency factor smaller (or larger) than the smallest (or largest) canonical efficiency factor. Any contrast vector $\mathbf{d}$ can be written as $\mathbf{d} = \sum a_i \mathbf{p}_i$ for some constants $a_1, a_2, \ldots, a_{v-1}$. Then

$$\text{var}(\mathbf{d}'\hat{\boldsymbol{\tau}}) = (\sigma^2/r)\sum(a_i^2/e_i)$$

Since for a complete block design $e_i = 1$ $(i = 1, 2, \ldots, v-1)$, the efficiency factor of the contrast $\mathbf{d}'\boldsymbol{\tau}$ is

$$E(\mathbf{d}'\boldsymbol{\tau}) = \frac{\sum a_i^2}{\sum(a_i^2/e_i)}$$

Let $e_{\min}$ and $e_{\max}$ be the smallest and largest canonical efficiency factors respectively. Then, as

$$e_{\max}^{-1}\sum a_i^2 \leqslant \sum(a_i^2/e_i) \leqslant e_{\min}^{-1}\sum a_i^2$$

it follows that

$$e_{\min} \leqslant E(\mathbf{d}'\boldsymbol{\tau}) \leqslant e_{\max} \tag{2.10}$$

Canonical efficiency factors provide a useful summary of the properties of a design. They are particularly important if the basic contrasts correspond to the contrasts of interest in the experiment, as is frequently the case in the multiple replicate factorial designs discussed in Chapter 7. The use of these factors to define optimality criteria is discussed in the next section.

2.4 Optimality criteria

Canonical efficiency factors provide a measure of the amount of information available on basic treatment contrasts from within-block comparisons. Since comparisons made within blocks are usually more accurately made than those between blocks, the aim in selecting a design, assuming all treatments to be of equal interest, should be to choose one which has these efficiency factors as large as possible. Various optimality criteria have been proposed.

If the harmonic mean of the canonical efficiency factors, or the average efficiency factor of pairwise treatment factors, of a design is at least as large as that of any other design then the design is said to be *A-optimal.* In view of (2.10), a criterion based on the smallest canonical efficiency factor is of interest since it indicates the worst that a design can do in estimating treatment contrasts. A design whose smallest canonical efficiency factor is at least as large as that of any other design is said to be *E-optimal.* A further criterion uses the geometric mean (or product) of the canonical efficiency factors. A design whose geometric mean efficiency factor is at least as large as that of any other design is said to be *D-optimal.* D-optimality is a criterion which has proved useful in the context of regression analysis, where it has an interpretation in terms of its equivalence with minimizing the maximum variance of predicted responses; such an interpretation is less meaningful for block designs.

A design which is optimal under any one of the above criteria is not necessarily optimal under the others; although John and Williams (1982) have conjectured that if a block design is A-optimal then it is also D-optimal. However, evidence gained from studies of different types of designs suggests that a design which is optimal or performs well on one criterion tends to perform well on the other criteria.

Another useful criterion is that of *(M, S)-optimality*, introduced by Shah (1960) and Eccleston and Hedayat (1974). It is a two-stage criterion. Firstly, a class of designs which maximizes the mean of the efficiency factors is formed (M-optimality). Then, within this class, the design (or designs) which minimize the spread or variance of the efficiency factors are identified (S-optimality). That is, $\sum e_i$ is first maximized and then $\sum e_i^2$ minimized. This criterion, therefore, uses the intuitively appealing requirement that all the efficiency factors should be as equal as possible.

If the canonical efficiency factors of a design are all equal then the design is clearly optimal under all four optimality criteria given above. Such a design is said to be *efficiency-balanced*; a complete block design is an example of such a design. From (2.4) it follows that for such designs

$$\mathbf{A} = rE \sum_{i=1}^{v-1} \mathbf{p}_i \mathbf{p}_i'$$

Now $\sum_{i=1}^{v} \mathbf{p}_i \mathbf{p}_i' = \mathbf{I}$, as it is the sum of v mutually orthogonal idempotent matrices, so that

$$\mathbf{A} = rE(\mathbf{I}_v - \mathbf{K}_v) \tag{2.11}$$

since $\mathbf{p}_v = (v^{-1/2})\mathbf{1}$. This matrix has only one distinct non-zero eigenvalue so that a design is efficiency-balanced if and only if the information matrix is of the form given by (2.11).

If $\mathbf{A}$ is given by (2.11) the design will also be *variance-balanced* in the sense that the variances of all estimated pairwise treatment differences will be the same, and equal to

$$\text{var}(\hat{\tau}_i - \hat{\tau}_j) = \frac{2\sigma^2}{rE} \qquad \text{for all } i \neq j \tag{2.12}$$

Although for equal replicate designs a variance-balanced design is also efficiency-balanced, Williams (1975b) shows that this result is not true for unequally replicated designs. This point is discussed further in Section 2.7.

The above criteria are appropriate when all treatments are of equal interest. However, experiments are often carried out in which some treatment contrasts are of more importance than others. Different criteria will, therefore, be needed for choosing designs appropriate for such experiments. Pearce (1974) proposed minimizing the weighted mean of the efficiency factors of interest. Freeman (1976a) suggested minimizing the weighted mean given by $\sum w_i \text{var}(\mathbf{c}_i'\hat{\boldsymbol{\tau}})$ where $\mathbf{c}_i'\boldsymbol{\tau}$ represents a contrast of interest and w_i is the weight to be attached to this contrast ($i = 1, 2, \ldots, t$, say). If $\mathbf{C}$ is the $v \times t$ matrix whose ith column is $\mathbf{c}_i$, and $\mathbf{w}$ is the weighting vector with ith element w_i, then the criterion is one of minimizing

$$\text{trace}(\mathbf{w}^\delta \mathbf{C}'\boldsymbol{\Omega}\mathbf{C}) = \text{trace}(\boldsymbol{\Omega}\mathbf{C}\mathbf{w}^\delta \mathbf{C}') \tag{2.13}$$

The difficulty, of course, is choosing appropriate weights as the choice in practice is most likely to be a highly subjective one. Jones (1976) has given details of a computer algorithm to derive optimal block designs using the criterion given in (2.13).

2.5 Simple counting rules for producing near-optimal designs

The search for optimal designs in a large family of designs of a given size invariable involves a mixture of theory and computing. The use of theoretical results can often limit the number of possible designs that have to be considered. Even so, there will often be large numbers of designs for which the canonical efficiency factors have to be computed. Eigenvalue computations are relatively time consuming, and consequently, attempts have been made to find criteria based on quick and simple counting rules. The idea is to use such rules to identify a small group of designs which are near optimal and then, if necessary, to choose the best designs from this group using the canonical efficiency factors and appropriate optimality criteria. That is, the counting rules are aimed at an initial sifting out of the less efficient designs.

One such rule is, in fact, given by the (M, S)-optimality criterion given in the previous section. Consider first the class of binary designs with $k \leqslant v$, defined in Section 1.6. Using (2.7),

$$\sum e_i = \text{trace}(\mathbf{A}^*) = v(k-1)/k$$

and

$$\sum e_i^2 = \text{trace}(\mathbf{A}^{*2}) = v - 2vr/k + (vr^2 + \sum_{i \neq j}\sum \lambda_{ij}^2)/k^2$$

where λ_{ij} gives the number of blocks containing both the ith and jth treatments. Hence, for designs of a given size, $\sum e_i$ is a constant and $\sum e_i^2$ is minimized by minimizing the sum of squares, $\sum\sum \lambda_{ij}^2$, of the off-diagonal elements of the concurrence matrix $\mathbf{NN}'$. An (M, S)-optimal binary design is, therefore, obtained by ensuring that the treatment concurrences are as equal as possible. The rule is simple to apply and does not involve the calculation of eigenvalues. However, it still may lead to a large class of designs. Further searching can be restricted to this class. For equal replicate designs, it has been conjectured that the A-, D- and E-optimal designs are to be found in the class of (M, S)-optimal designs; see John and Williams (1982).

When $k > v$, the designs are necessarily non-binary. An (M, S)-optimal non-binary design belongs to the class of designs obtained by appending a binary design to sets of complete block designs. That is, the required design will have incidence matrix $\mathbf{N} = p\mathbf{J} + \mathbf{N}_0$ where p is some positive integer and $\mathbf{N}_0$ is the incidence matrix of the binary design. Further, if the binary design is (M, S)-optimal then so is the corresponding non-binary design.

Although minimizing trace$[(\mathbf{NN}')^2]$ is a quick and simple method of finding (M, S)-optimal designs, it will in general still produce a large class of designs. Further, not all these designs will be good on other optimality criteria. It becomes necessary, therefore, to apply other rules to pick out the best designs from this class. Lamacraft and Hall (1982), in constructing tables of cyclic designs, used, at a second stage, the criterion of minimizing trace$[(\mathbf{NN}')^3]$. A rationale behind this choice, for connected designs, is as follows. The non-singular matrix $\mathbf{A}^* + \mathbf{J}/rk$, where $\mathbf{A}^*$ is given in (2.7), has one eigenvalue equal to $v^{1/2}/rk$ corresponding to the eigenvector $\mathbf{p}_v = (v^{-1/2})\mathbf{1}$ with the remaining eigenvalues being equal to the canonical efficiency factors. Since, for some constant α,

$$(\mathbf{A}^* + \mathbf{J}/rk)^{-1} = \mathbf{I} + \alpha\mathbf{J} + \sum_{i=1}^{\infty} (rk)^{-i}(\mathbf{NN}')^i$$

it follows that

$$\sum_{j=1}^{v-1} e_j^{-1} = \text{constant} + \sum_{i=2}^{\infty} (rk)^{-i}\text{trace}(\mathbf{NN}')^i \qquad (2.14)$$

The traces of successive powers of $\mathbf{NN}'$ provide, therefore, simple criteria to use in the search for A-optimal designs. The smaller powers make the greatest contribution to the sum in (2.14), so that it becomes intuitively reasonable to minimize sequentially trace$[(\mathbf{NN}')^2]$, trace$[(\mathbf{NN}')^3]$, and so on, in the search for efficient designs.

A similar procedure has been proposed by Paterson (1983). He suggested the sequential use of the number of circuits of different length in the treatment concurrence graph of a design. This graph has been defined in Section 1.6.

Example 2.1
Consider the following design for 6 treatments labelled 0, 1,..., 5, in 9 blocks of 2, where blocks are set out in columns:

0	1	2	3	4	5	0	1	2
1	2	3	4	5	0	3	4	5

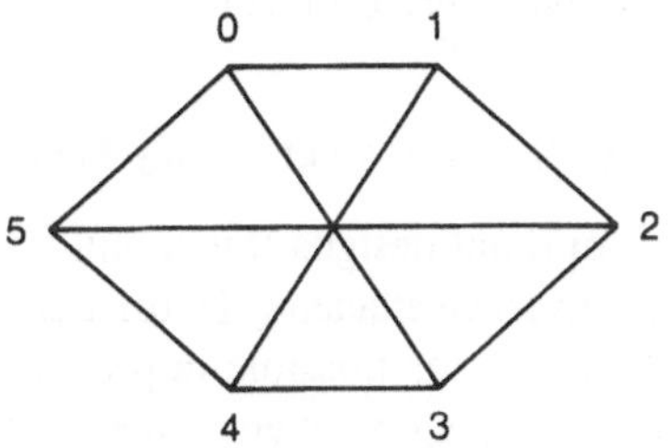

Figure 2.1 *The treatment concurrence graph for Example 2.1.*

Its treatment concurrence graph is shown in Fig. 2.1. Treatment 0, for instance, occurs in the same block as treatments 1, 3 and 5 but not with 2 and 4. Hence lines connect point 0 with points 1, 3 and 5 but not with points 2 and 4.

A *path* joining points i and j is a route from i to j traversing lines in the graph, whereas a *circuit* is a path joining a point to itself. For example, in the graph in Fig. 2.1, 0–1–2 is a path joining points 0 and 2, and 0–1–4–5–0 is a circuit at point 0.

Each path in the graph contributes to the estimate of a pairwise treatment difference in the design. To illustrate this, suppose that the observations in the first two blocks in the design of Exmple 2.1 are y_1, y_2, y_3 and y_4 respectively. Then, assuming the intra-block model (1.1), $y_1 - y_2$ provides an estimate of $\tau_0 - \tau_1$ with variance $2\sigma^2$, and $(y_1 - y_2) - (y_3 - y_4)$ an estimate of $\tau_0 - \tau_2$ with variance $4\sigma^2$. The first estimate corresponds to the path 0–1 of length 1, while the second corresponds to the path 0–1–2 of length 2. In general, a path of length h will contribute an estimate of a treatment difference with variance $2h\sigma^2$. Paths of differing lengths give estimates which can be combined to give overall estimates of each pairwise treatment difference.

Circuits, on the other hand, contribute nothing useful to the estimation of such treatment differences, since a comparison of a treatment with itself is of no interest. Consequently, good designs might be expected to have few circuits of short lengths. If c_h is the number of circuits of length h in the treatment concurrence graph, then Paterson (1983) argues that a reasonable criterion to use in the search for good designs would be to sequentially minimize c_2, c_3, c_4 and so on.

The fact that $c_h = \text{trace}(\mathbf{B}^h)$, where $\mathbf{B}$ is the adjacency matrix of the treatment concurrence graph given by $\mathbf{B} = \mathbf{NN}' - r\mathbf{I}$, connects this approach with that based on (2.14).

2.6 Upper bounds for the average efficiency factor

In the search for A-optimal designs it is useful to have available an upper bound for the average efficiency factor E as a measure against which to assess the scope for possible improvement. For instance, a computer search could stop when a design is found which is sufficiently close to the bound. A number of upper bounds have been proposed and, for a given set of parameters, the best bound will be provided by the minimum value of those available.

Since the canonical efficiency factors are all positive, the harmonic mean cannot exceed the arithmetic mean $\bar{e}$ of these factors. Hence, the simplest upper bound for the average efficiency factor E is $\bar{e} = \text{trace}(\mathbf{A}^*)/(v-1)$, where $\mathbf{A}^*$ is given by (2.7). For binary designs, $\text{trace}(\mathbf{A}^*) = v(k-1)/k$, so that the bound is

$$U_0 = \frac{v(k-1)}{k(v-1)} \tag{2.15}$$

For non-binary designs the trace($\mathbf{A}^*$) may vary from one design to another. By finding a lower bound for the trace of the matrix $\mathbf{NN}'$, Williams, Patterson and John (1976) show that an improved upper bound is given by

$$U_1 = 1 - \frac{k'(v-k')}{k^2(v-1)} \tag{2.16}$$

where $k' = k \bmod v$. For binary designs $U_1 = U_0$ but for non-binary designs $U_1 \leqslant U_0$.

In general, $E = U_0$ only when the canonical efficiency factors of the design are all equal, i.e. when the design is efficiency-balanced and the information matrix is of the form given by (2.11). Improved bounds can be obtained for those cases where efficiency-balanced designs cannot be constructed. Conniffe and Stone (1974) give an upper bound for binary designs which has been extended by Jarrett (1977) to cover any block design. The bound is

$$U_2 = U_0 - \frac{(v-2)S^2}{U_0 + (v-3)S} \tag{2.17}$$

where

$$(v-1)(v-2)S^2 = S_2 = \sum (e_i - \bar{e})^2 \tag{2.18}$$

If the design is efficiency-balanced then $S^2 = 0$ and $U_2 = U_0$.

The bound U_2 requires knowledge of S^2. However, for equal replicate and equal block size binary designs it is possible to find a lower bound for S^2 which then leads to an upper bound for E based solely on the design parameters; note that U_2 is a decreasing function of S^2. It can be shown that

$$r^2k^2S_2 = \sum_{i \neq j}\sum (\lambda_{ij} - \bar{\lambda})^2$$

where λ_{ij} is the (ij)th element of $\mathbf{NN}'$ and $\bar{\lambda} = r(k-1)/(v-1)$ is the mean of the λ_{ij} over all $i \neq j$. Hence, a lower bound for S^2 for binary designs occurs when the off-diagonal elements of $\mathbf{NN}'$ differ by at most one. This gives

$$S^2 \geqslant S_L^2 = \frac{v\alpha(1-\alpha)}{(v-2)r^2k^2} \tag{2.19}$$

where α is the fractional part of $\bar{\lambda}$. The upper bound for E is now given by U_2 with S^2 replaced by S_L^2.

Essentially U_2 is making use of the information available on the number of circuits of length 2, c_2, in the treatment concurrence graph of the design; it can be shown that S_2 is a monotonic function of c_2. It follows in the light of the discussion in the previous section that tighter bounds might be expected if information on the number of circuits of lengths 3, 4, ... were used. Jarrett (1983) gives a bound which uses the number of triangles c_3. Let S_2 be defined as in (2.18) and let $S_3 = \sum(e_i - \bar{e})^3$, then the bound is

$$U_3 = U_0 - \frac{S_2^2}{(v-1)(S_3 + U_0 S_2)} \tag{2.20}$$

This bound requires knowledge of both S_2 and S_3. Jarrett (1983) substitutes bounds for S_2 and S_3 into U_3 to give an upper bound in terms of the design parameters for 2-concurrence designs, i.e. designs for which each of the off-diagonal elements of $\mathbf{NN}'$ takes one of two possible values. Of particular interest is the case where the off-diagonal elements differ by one. In the absence of efficiency-balanced designs the class of such designs are, of course, (M, S)-optimal and have been conjectured to include the A-optimal designs; see Section 2.5. With this constraint, substitution of bounds for S_2 and S_3 into U_3 leads to

$$U_4 = U_0 - \frac{v\alpha(1-\alpha)}{rk\{rkU_0 + z - (v+1)\alpha\}} \tag{2.21}$$

where

$$z = \begin{cases} v, & \text{if } 2(v-1)\alpha \geqslant v \\ (v-2)\alpha/(1-\alpha), & \text{if } 2(v-1)\alpha < v \end{cases}$$

The bound U_4 is conjectured to be an upper bound for E. Jarrett (1983) showed that U_4 is a tighter upper bound than the one given earlier by Williams and Patterson (1977), which also used the constraint that the off-diagonal elements of $\mathbf{NN}'$ do not differ by more than one.

The upper bounds given above are best used when $v \leqslant b$. When $v > b$, better bounds can be obtained if use is made of the relationships which exist between a design and its dual. This will be considered in Section 2.8.

2.7 Efficiency factors for unequally replicated designs

For designs with unequal treatment replication, two measures of efficiency factors can be defined. Which should be used to assess different designs will depend on the constraints imposed by experimental requirements.

One measure will be appropriate when the replication vector is regarded as fixed. That is, it is a condition of the experiment that the ith treatment is replicated r_i times in the design, where r_i is fixed $(i = 1, 2, \ldots, v)$. Efficiency factors are then obtained by comparing variances of estimated treatment contrasts with those obtained from an orthogonal design, i.e. with

$$\operatorname{var}(\mathbf{c}'\hat{\mathbf{t}}) = \mathbf{c}'\mathbf{r}^{-\delta}\mathbf{c}\sigma^2 \tag{2.22}$$

This provides a measure of the effectiveness of blocking since (2.22) gives the variances that would have been obtained (assuming the same σ^2) if blocks had been ignored in the design.

The second measure will be more suitable when the total number of experimental units is fixed, i.e. when $n = \mathbf{r}'\mathbf{1}$ is fixed but not the individual elements of the vector $\mathbf{r}$. A basis for the comparison of designs will then be the average replication $\bar{r} = n/v$. Hence, for connected designs, if $\lambda_1, \lambda_2, \ldots, \lambda_{v-1}$ are the non-zero eigenvalues of the information matrix $\mathbf{A}$ given in (1.12) then, following (2.6), corresponding efficiency factors are given by $\lambda_j/\bar{r}$ $(j = 1, 2, \ldots, v-1)$. Criteria based on these efficiency factors can then be used to assess different designs. In general, when treatments are of equal interest, optimal designs will have the property that the treatments are equally replicated; Jones and Eccleston (1980), however, give three examples with block size 2 where the A-optimal designs have unequal rather than equal replication. Of course, an equally replicated design will not be available if $\bar{r}$ is not an integer.

Since this second measure is relatively straightforward, the remainder of this section will be concerned with the case where $\mathbf{r}$ is regarded as fixed. Following Ceranka and Mejza (1979), the concepts of a basic contrast and a canonical efficiency factor can be generalized. Let e_j be an eigenvalue corresponding to an eigenvector $\mathbf{s}_j$ of the matrix $\mathbf{A}$ given in (1.12) with respect to $\mathbf{r}^\delta$, i.e. $\mathbf{A}\mathbf{s}_j = e_j\mathbf{r}^\delta\mathbf{s}_j$ $(j = 1, 2, \ldots, v)$. The eigenvectors $\mathbf{s}_j$ can be chosen to satisfy

$$\mathbf{s}_i\mathbf{r}^\delta\mathbf{s}_j = \begin{cases} 1, & i = j \\ 0, & i \neq j \end{cases} \tag{2.23}$$

Again, since $\mathbf{A}$ is a singular matrix, at least one eigenvalue, e_v say, is zero with corresponding eigenvector $\mathbf{s}_v = (n^{-1/2})\mathbf{1}$. For a connected design all other $v - 1$ eigenvalues are non-zero. Now if $\mathbf{p}_j = r^\delta\mathbf{s}_j$ $(j = 1, 2, \ldots, v - 1)$ then $\mathbf{p}_j'\boldsymbol{\tau}$ is a contrast in the treatment parameters. Further, using (1.23) and the above results,

$$\begin{aligned}
\operatorname{var}(\mathbf{p}_j'\hat{\boldsymbol{\tau}}) &= \mathbf{p}_j'\boldsymbol{\Omega}\mathbf{p}_j\sigma^2 \\
&= \mathbf{s}_j'\mathbf{r}^\delta\boldsymbol{\Omega}\mathbf{r}^\delta\mathbf{s}_j\sigma^2 \\
&= e_j^{-2}\mathbf{s}_j'\mathbf{A}\boldsymbol{\Omega}\mathbf{A}\mathbf{s}_j\sigma^2 \\
&= e_j^{-2}\mathbf{s}_j'\mathbf{A}\mathbf{s}_j\sigma^2 \\
&= e_j^{-1}\sigma^2
\end{aligned}$$

For a complete block design, with $\boldsymbol{\Omega} = \mathbf{r}^{-\delta}$,

$$\operatorname{var}(\mathbf{p}_j'\hat{\boldsymbol{\tau}}) = \mathbf{s}_j'\mathbf{r}^\delta\mathbf{r}^{-\delta}\mathbf{r}^\delta\mathbf{s}_j = \mathbf{s}_j'\mathbf{r}^\delta\mathbf{s}_j\sigma^2 = \sigma^2$$

Hence, the efficiency factor of the contrast $\mathbf{p}_j'\boldsymbol{\tau}$ is e_j. These contrasts are the basic contrasts of the design and the e_j are the canonical efficiency factors. Extending (2.7) to the unequal replicate case, the canonical efficiency factors are more conveniently obtained as the eigenvalues of the symmetric matrix

$$\mathbf{A}^* = \mathbf{r}^{-\delta/2}\mathbf{A}\mathbf{r}^{-\delta/2} = \mathbf{I} - r^{-\delta/2}\mathbf{N}\mathbf{k}^{-\delta}\mathbf{N}'\mathbf{r}^{-\delta/2} \tag{2.24}$$

where $\mathbf{r}^{-\delta/2}$ is the inverse of $\mathbf{r}^{\delta/2}$, which is the diagonal matrix with ith diagonal element equal to $r_i^{1/2}$.

With unequal replicate designs, therefore, optimality criteria based on the eigenvalues of the information matrix $\mathbf{A}$ or the matrix $\mathbf{A}^*$ in (2.24) can be used depending on whether the purpose is to compare designs where only the total size of the experiment is fixed or where the replication vector is fixed.

2.8 Duality

The dual design of a block .design with incidence matrix $\mathbf{N}$ is the block design with incidence matrix $\mathbf{N}'$. Thus the dual is obtained by interchanging the treatment and block symbols in the original design.

Example 2.2
Consider the following binary design for $v = 6, r = 2, k = 4$ and $b = 3$:

$$\begin{array}{cccc}(0 & 1 & 3 & 4)\\(0 & 2 & 3 & 5)\\(1 & 2 & 4 & 5)\end{array}$$

The incidence matrix $\mathbf{N}$, with a row for each treatment and a column for each block, is

$$\mathbf{N} = \begin{pmatrix} 1 & 1 & 0 \\ 1 & 0 & 1 \\ 0 & 1 & 1 \\ 1 & 1 & 0 \\ 1 & 0 & 1 \\ 0 & 1 & 1 \end{pmatrix}$$

Consequently, the incidence matrix of the dual design is

$$\mathbf{N}' = \begin{pmatrix} 1 & 1 & 0 & 1 & 1 & 0 \\ 1 & 0 & 1 & 1 & 0 & 1 \\ 0 & 1 & 1 & 0 & 1 & 1 \end{pmatrix}$$

Hence, the dual design is

$$(0\ \ 1)\quad (0\ \ 2)\quad (1\ \ 2)\quad (0\ \ 1)\quad (0\ \ 2)\quad (1\ \ 2)$$

and has parameters $v_d = b = 3$, $r_d = k = 4$, $k_d = r = 2$ and $b_d = v = 6$.

Two useful relationships between a design and its dual can now be established. One concerns the canonical efficiency factors and their harmonic mean, and the other the generalized inverse of the information matrix. It will be assumed, for the first relationship, that the original design and, hence, its dual design are connected.

The average efficiency factor E_d, say, of the dual design is the harmonic mean of the non-zero eigenvalues of the $b \times b$ matrix

$$\mathbf{A}_d^* = \mathbf{I} - \mathbf{k}^{-\delta/2}\mathbf{N}'\mathbf{r}^{-\delta}\mathbf{N}\mathbf{k}^{-\delta/2} \tag{2.25}$$

This follows from the form of A^* given in (2.24). Note that

$$\mathbf{A}^* = \mathbf{I} - \mathbf{B}\mathbf{B}', \qquad \mathbf{A}_d^* = \mathbf{I} - \mathbf{B}'\mathbf{B}$$

where $\mathbf{B} = \mathbf{r}^{-\delta/2}\mathbf{N}\mathbf{k}^{-\delta/2}$.

Now consider the case where $v > b$. $\mathbf{A}_d^*$ then has a single zero eigenvalue and $(b-1)$ non-zero eigenvalues $e_1, e_2, \ldots, e_{b-1}$. Then

$$E_d = \frac{b-1}{\sum_{i=1}^{b-1} e_i^{-1}} \tag{2.26}$$

As the matrices $\mathbf{BB}'$ and $\mathbf{B}'\mathbf{B}$ have the same non-zero eigenvalues, the eigenvalues of the matrix $\mathbf{A}^*$ of (2.24) consist of the b eigenvalues of $\mathbf{A}_d^*$ and $(v-b)$ eigenvalues equal to 1. Hence, the average efficiency factor of the original design is

$$E = \frac{v-1}{(v-b) + \sum_{i=1}^{b-1} e_i^{-1}}$$

Using (2.26) gives

$$E = \frac{v-1}{(v-b) + (b-1)E_d^{-1}} \tag{2.27}$$

If $v \leqslant b$, then $\mathbf{A}_d^*$ has a single zero eigenvalue, $(v-1)$ non-zero eigenvalues $e_1, e_2, \ldots, e_{v-1}$, and $(b-v)$ eigenvalues equal to 1. Hence

$$E_d = \frac{b-1}{\sum_{i=1}^{v-1} e_i^{-1} + (b-v)} \tag{2.28}$$

Since the non-zero eigenvalues of $\mathbf{A}^*$ are now $e_1, e_2, \ldots, e_{v-1}$, it follows from (2.28) that E is again given by (2.27). This establishes a relationship between the average efficiency factors of a design and its dual. Note that E increases with E_d.

These results are useful in a number of ways. If v is very much greater than b then the canonical efficiency factors and the average efficiency factor E are most easily obtained from the dual design. If it is known that a design has a high E factor then, from (2.27), so will its dual. For example, for the dual design with three treatments in Example 2.2 it is readily shown that $\mathbf{A}_d^*$ has the form (2.11) so that the design is efficiency-balanced with $E_d = 3/4$. It follows that the design with 6 treatments in blocks of 4 is A-optimal with $E = 15/17$, using (2.27). This design is also D- and E-optimal.

When $v > b$, better upper bounds for E can be obtained through the dual design than by direct use of the quantities given in Section 2.6. An appropriate upper bound is first obtained for the dual design using either U_0, U_1, U_2 or U_4. The chosen bound is then used in (2.27) in place of E_d.

Example 2.3

Suppose an upper bound on E is required for designs with $v = 12$, $k = 4$, $r = 2$ and $b = 6$. Direct use of (2.21) gives $U_4 = 0.786$. For the dual design with $v_d = 6$, $k_d = 2$, $r_d = 4$ and $b_d = 12$ the bound is $U_4 = 0.577$. Substituting this value for E_d in (2.27) leads to the bound 0.750, which is considerably less than $U_4 = 0.786$.

The second relationship between a design and its dual concerns the generalized inverse of the information matrix. For the dual design the information matrix becomes

$$\mathbf{A}_d = \mathbf{k}^\delta - \mathbf{N}'\mathbf{r}^{-\delta}\mathbf{N} \tag{2.29}$$

If $\mathbf{\Omega}_d$ is a generalized inverse of this matrix then it can be established that

$$\mathbf{\Omega} = \mathbf{r}^{-\delta}[\mathbf{r}^\delta + \mathbf{N}\mathbf{\Omega}_d\mathbf{N}']\mathbf{r}^{-\delta} \tag{2.30}$$

is a generalized inverse of the information matrix $\mathbf{A}$ by showing that $\mathbf{A\Omega A} = \mathbf{A}$.

If $\mathbf{\Omega}_d$ is easier to obtain than $\mathbf{\Omega}$, which would often be the case, for instance, if v is very much greater than b, then $\mathbf{\Omega}$ can be conveniently obtained by substituting $\mathbf{\Omega}_d$ into (2.30).

CHAPTER 3

Traditional block designs

3.1 Introduction

The next two chapters will be concerned with block designs with equal block size k and in which each of the v treatments is replicated r times. Orthogonality between treatments and blocks will result only if the designs are *complete block* designs, i.e. have each treatment occurring equally frequently in each block. In many experimental situations when the number of treatments is large or the amount of homogeneous material available is restricted, it is often not feasible to have even a single replicate of every treatment in each block. In such cases it becomes necessary to use designs in which the block size is less than the number of treatments, i.e. $k < v$. These designs are called *incomplete block* designs.

In 1936 Yates proposed the use of *balanced incomplete block* and *lattice* designs (Yates, 1936a, b). Balanced incomplete block designs have the property that comparisons between pairs of treatments are all made with the same accuracy; they are, in fact, efficiency-balanced binary designs (see Section 2.4). Lattice designs are resolvable in the sense that the blocks can be grouped in such a way that each treatment is replicated once in every group. Comparisons between pairs of treatments now vary in accuracy, but fewer replicates are required than for many of the balanced designs. From a consideration of their dual designs, the properties of lattice designs are readily obtained.

Balanced incomplete block and lattice designs only exist, however, for a limited number of combinations of v, r and k. When designs of the required size are not available, other incomplete block designs will be required. At the time when alternatives to balanced incomplete block and lattice designs were being sought, simple desk calculators were being used for the statistical analysis of experimental data. Consequently, the emphasis was on obtaining designs or classes of designs which allowed simple analyses and, in particular, easily obtainable variance–covariance matrices. The mathematical and combinatorial properties, rather than the statistical properties, of designs were, therefore, of primary interest.

In 1939 Bose and Nair introduced the class of *partially balanced incomplete block designs with m associate classes* (PBIB/m designs). The designs satisfy the following requirements:

1. There are v treatments in b blocks of k units per block with each treatment replicated r times.
2. Any two treatments are either 1st, 2nd,…, mth associates.
3. Each treatment has exactly n_i ith associates ($i = 1,\dots,m$).
4. Given any two treatments which are ith associates, the number of treatments common to the jth associates of the first, and the kth associates of the second is p^i_{jk} and is independent of the two treatments chosen. Also $p^i_{jk} = p^i_{kj}$ ($i, j, k = 1,\dots,m$).
5. Two treatments which are ith associates occur together in exactly λ_i blocks ($i = 1,\dots,m$).

This definition of a PBIB/m design ensures that the $\mathbf{\Omega}$ matrix of (1.37) has equal diagonal elements and m distinct off-diagonal elements (Shah, 1959). A PBIB/m design, therefore, has m distinct pairwise efficiency factors. Further, the information matrix $\mathbf{A}$ has at most m distinct non-zero eigenvalues, so that the designs have at most m distinct canonical efficiency factors. Balanced incomplete block designs are a special case within this class with $m = 1$.

The simplest class of partially balanced designs, apart from balanced incomplete blocks, are those with $m = 2$ associate classes. Since 1939, PBIB/2 designs have been extensively studied. The important concept of an association scheme was developed by Bose and Shimamoto (1952) and formed the basis for a classification of PBIB/2 designs. A catalogue of designs was published by Bose, Clatworthy and Shrikhande (1954) and this has been extensively revised and extended by Clatworthy (1973). He classifies PBIB/2 designs into six types, namely:

1. group divisible
2. triangular
3. Latin square types
4. cyclic
5. partial geometry
6. miscellaneous

The most important type are the group-divisible designs, with almost three-fifths of the designs in Clatworthy's catalogue of this type. An important feature of these designs is that the treatments are divided into groups of equal size, and such a division will often correspond to the way an experimenter will subdivide a collection of treatments. Further, the designs are useful in factorial experiments, as will be shown in Chapter 7.

The other types of PBIB/2 designs do provide many efficient incomplete block designs. However, a very large number are relatively inefficient and are unlikely to be of much practical value. In most of these cases more efficient designs can be obtained by other methods. Only group-divisible designs will, consequently, be discussed in this book. Excellent accounts of PBIB/2 designs can, however, be found in Clatworthy (1973) and in the books by P.W.M. John (1971) and Raghavarao (1971).

Even with the addition of PBIB/2 designs, there are still many parameter combinations for which there is no design tabulated or where the design is inefficient. Little work has been carried out into PBIB/m designs with $m > 2$, mainly because of the difficulties involved in constructing such designs. In any case, with the advent of the computer there was not the same need for designs with simple methods of analysis.

Other classes of incomplete block designs have been obtained using methods of construction which are relatively simple but which are capable of producing a large number of designs. The efficient designs of a given size within a class can then be identified, using a combination of theoretical results and computing. *Cyclic designs* provide an important and extensive class of such designs.

This chapter will be concerned with what might be called the traditional block designs, namely complete blocks, balanced incomplete blocks, lattices and group-divisible designs. Cyclic designs and variants based on the cyclical method of construction will be discussed in detail in the next chapter.

All the incomplete block designs considered will be binary, i.e. each treatment will occur at most once in each block. Non-binary designs could, of course, be constructed with $k < v$. However, they will generally be less efficient than the binary designs. The sum of the canonical efficiency factors is clearly maximized for binary designs; this being the same as minimizing the trace of the matrix $\mathbf{NN}'$ or minimizing the sum of squares of the elements of the matrix $\mathbf{N}$.

3.2 Complete block designs

In a complete block design each treatment occurs equally frequently, p times say, in every block. This means that the number of units per block must be a multiple of the number of treatments, and the number of replicates must be a multiple of the number of blocks, i.e.

$$k = pv, \qquad r = pb \tag{3.1}$$

The most widely used designs in practice are those with $p = 1$; the so called *randomized block* designs.

Many of the properties of complete block designs have already been given in Section 1.7. Estimates of treatment parameters are based on treatment means with no adjustment for blocks being necessary, as blocks and treatments are orthogonal. Hence, the contrast $\mathbf{c}'\boldsymbol{\tau} = \sum c_i \tau_i$ in the treatment parameters is estimated by $\sum c_i \bar{y}_i$ with variance $(\sigma^2/r)\sum c_i^2$, where $\bar{y}_i$ is the ith treatment mean.

The information matrix $\mathbf{A}$ is, from (1.40), given by

$$\mathbf{A} = r(\mathbf{I}_v - \mathbf{K}_v) \tag{3.2}$$

so that, comparing with (2.11), complete block designs are efficiency-balanced with all efficiency factors equal to unity. Thus, all the information on treatment contrasts is available from comparisons made within blocks. The analysis of variance for testing for differences between treatment or block effects is given in Table 3.1.

Since any contrast vector is an eigenvector of $\mathbf{A}$ given in (3.2), the basic contrasts of a complete block design are given by any set of $(v-1)$ orthogonal contrasts. Following from (2.8), this means that the treatment sum of squares in Table 3.1 can be written as

$$S(T) = \sum_{i=1}^{v-1} (\mathbf{c}_i'\mathbf{T})^2/(r\mathbf{c}_i'\mathbf{c}_i)$$

Table 3.1 *Analysis of variance for a complete block design.*

	d.f.	*s.s.*
Between blocks	$b-1$	$(1/k)\mathbf{B}'\mathbf{B} - G^2/n$
Between treatments	$v-1$	$(1/r)\mathbf{T}'\mathbf{T} - G^2/n$
Residual	$n-b-v+1$	by subtraction
Total	$n-1$	$\mathbf{y}'\mathbf{y} - G^2/n$

for any set of contrast vectors $\mathbf{c}_i$ satisfying $\mathbf{c}_i'\mathbf{c}_j = 0$ $(i, j = 1, 2, \ldots, v-1; i \neq j)$. The $v-1$ degrees of freedom for the treatment sum of squares can, therefore, be partitioned into $v-1$ single degree of freedom components by using a set of $v-1$ orthogonal contrasts. The partition is not unique and a choice of an appropriate set of contrasts depends on the purpose of the experiment.

Example 3.1

For an experiment with four treatments consider the three sets of contrasts in Table 3.2. Set (a) might be appropriate if, for example, treatment 0 was a control treatment and the other treatments were test treatments. The first contrast compares the control treatment with the test treatments, while the other two contrasts involve comparisons among the test treatments. Set (b) would be appropriate if, for example, the four treatments were made up of combinations of two factors each at two levels. The three contrasts would then represent the main effects and interaction of these factors; this is an example of a factorial experiment. Finally, set (c) corresponds to the linear, quadratic and cubic components of four equally spaced quantitative treatment factors.

Table 3.2 *Sets of contrasts for four treatments.*

	Treatment			
Set	*0*	*1*	*2*	*3*
	3	−1	−1	−1
(a)	0	1	0	−1
	0	−1	2	−1
	1	1	−1	−1
(b)	1	−1	1	−1
	1	−1	−1	1
	−3	−1	1	3
(c)	1	−1	−1	1
	−1	3	−3	1

3.3 Balanced incomplete block designs

Balanced incomplete block designs are binary designs in which every pair of treatments occurs together in exactly λ (say) blocks. For example, a balanced incomplete block design for $v = b = 5$ and $r = k = 4$ is

$$\begin{matrix} (0 & 1 & 2 & 3) \\ (0 & 1 & 2 & 4) \\ (0 & 1 & 3 & 4) \\ (0 & 2 & 3 & 4) \\ (1 & 2 & 3 & 4) \end{matrix}$$

Each pair of treatments can be seen to occur together in exactly three blocks, so that $\lambda = 3$.

The parameters v, k, r, b and λ of the balanced incomplete block design satisfy the two relationships

$$bk = vr \tag{3.3}$$

and

$$r(k-1) = \lambda(v-1) \tag{3.4}$$

Each side of (3.3) represents the total number of experimental units or plots and is, in fact, a relationship that holds for all block designs. Equation (3.4) is established by noting that a given treatment will occur with $(k-1)$ other treatments in each of r blocks and also occurs with each of the other $(v-1)$ treatments in λ blocks.

A balanced incomplete block design cannot exist unless both (3.3) and (3.4) are satisfied. For instance, no design will exist for $v = b = 6$ and $r = k = 3$ since, from (3.4), $\lambda = 6/5$ is not an integer. However, these conditions are not sufficient for the existence of a balanced incomplete block design. Even if both (3.3) and (3.4) are satisfied it does not follow that such a design exists. For example, no balanced incomplete block design exists when $v = 15$, $r = 7$, $k = 5$, $b = 21$ and $\lambda = 2$ even though both conditions are satisfied.

3.3.1 Construction of balanced incomplete block designs

The balanced incomplete block design for $v = 5$ treatments given above is an example of an *unreduced* design. Such designs are obtained by taking all combinations of the v treatments k at a time. An unreduced design would, therefore, require

$$b = \frac{v!}{k!(v-k)!} \quad \text{blocks and} \quad r = \frac{(v-1)!}{(k-1)!(v-k)!} \quad \text{replicates}$$

and will have

$$\lambda = \frac{(v-2)!}{(k-2)!(v-k)!}$$

Hence, unreduced designs usually require a large number of blocks and replicates so that the resulting designs will often be too large for practical purposes. Other balanced incomplete block designs can, however, be constructed which require fewer blocks and replicates than the unreduced designs. Two series of designs for $v = n^2$ and $v = n^2 + n + 1$ treatments respectively can be constructed from complete sets of $n \times n$ *mutually orthogonal Latin squares.*

A Latin square is an arrangement of n symbols or labels in an $n \times n$ array such that each label occurs once in each row and once in each column of the array. For example, the following are 4×4 Latin squares, where the labels are denoted by A, B, C and D:

```
A B C D    A B C D    A B C D
B A D C    C D A B    D C B A
C D A B    D C B A    B A D C
D C B A    B A D C    C D A B
```

Two Latin squares are pairwise orthogonal if, when one square is superimposed on the other, each label of one square occurs once with each label of the other square. Three or more squares are mutually orthogonal if they are pairwise orthogonal. The three 4×4 Latin squares above are mutually orthogonal.

For some values of n there will exist a *complete* set of $n - 1$ mutually orthogonal Latin squares. Complete sets are known to exist for any $n = p^s$, where p is a prime number. Tables can be found in Fisher and Yates (1963). The complete sets are used to construct two series of balanced incomplete block designs as follows.

Suppose $v = n^2$ treatments are set out in an $n \times n$ array. A group of n blocks, each of size n, is obtained by letting the rows of the array represent blocks. Another group of n blocks is given by taking the columns of the array as blocks. Now suppose one of the orthogonal squares is superimposed onto the array of treatments. A further group of n blocks will be obtained if all treatments common to a particular label in the square are placed in a block. Each of the $n-1$ orthogonal squares will produce a set of n blocks in this manner. The resulting design is a balanced incomplete block design with $v = n^2$, $k = n$, $r = n + 1$, $b = n(n + 1)$ and $\lambda = 1$.

Example 3.2

For $n = 3$, a complete set of orthogonal 3×3 squares and the $v = n^2 = 9$ treatments in a 3×3 array are:

A	B	C	A	B	C	0	1	2
C	A	B	B	C	A	3	4	5
B	C	A	C	A	B	6	7	8

Four groups of 3 blocks are obtained from the rows, columns and labels of the two squares, as follows:

(rows):	(0 1 2)	(columns):	(0 3 6)
	(3 4 5)		(1 4 7)
	(6 7 8)		(2 5 8)
(1st square):	(0 4 8)	(2nd square):	(0 5 7)
	(1 5 6)		(1 3 8)
	(2 3 7)		(2 4 6)

It can be checked that this a balanced incomplete block design with $v = 9$, $b = 12$, $k = 3$, $r = 4$ and $\lambda = 1$.

The *complement* of the design in Example 3.2, obtained by replacing treatments in a block by those not in the block, is also a balanced incomplete block. In general, the complementary design will be a balanced incomplete block design with $v = n^2$, $k = n(n-1)$, $r = n^2 - 1$, $b = n(n + 1)$ and $\lambda = n^2 - n - 1$.

The $n(n + 1)$ blocks of the designs for $v = n^2$ treatments have been arranged in $n + 1$ groups of n blocks each. Now suppose a new treatment is added to all the blocks in a particular group and that the treatment added is different for each group; also, that one further block is obtained which consists entirely of these $n + 1$ new treatments. This method produces a second series of balanced incomplete block designs with $v = b = n^2 + n + 1$, $r = k = n + 1$ and $\lambda = 1$. Its complement is also a balanced incomplete block design with the same v and b, and $r = k = n^2$, $\lambda = n(n-1)$.

Example 3.3
A balanced incomplete block design for $v = b = 13$ and $r = k = 4$, obtained from the design in Example 3.2 for $n = 3$, is given by

$$\begin{array}{cc} (0\ 1\ 2\ 9) & (0\ 3\ 6\ 10) \\ (3\ 4\ 5\ 9) & (1\ 4\ 7\ 10) \\ (6\ 7\ 8\ 9) & (2\ 5\ 8\ 10) \\ & \\ (0\ 4\ 8\ 11) & (0\ 5\ 7\ 12) \\ (1\ 5\ 6\ 11) & (1\ 3\ 8\ 12) \\ (2\ 3\ 7\ 11) & (2\ 4\ 6\ 12) \end{array}$$

$$(9\ 10\ 11\ 12)$$

Its complement has $r = k = 9$ and $\lambda = 6$.

Unreduced designs and designs obtained using orthogonal squares account for a large number of known balanced incomplete block designs. Various methods have been used to construct other balanced incomplete block designs and detailed accounts can be found in the books by John (1971) and Raghavarao (1971). Tables of balanced incomplete block designs are given in Fisher and Yates (1963).

3.3.2 *Properties of balanced incomplete block designs*

For a balanced incomplete block design, the diagonal elements of the concurrence matrix $\mathbf{NN}'$ are all equal to r and the off-diagonal elements are all equal to λ, i.e.

$$\mathbf{NN}' = (r - \lambda)\mathbf{I} + \lambda\mathbf{J}.$$

Hence, the information matrix $\mathbf{A}$ of (1.12) is

$$\mathbf{A} = rE(\mathbf{I} - \mathbf{K}) \tag{3.5}$$

where

$$E = \lambda v/rk \tag{3.6}$$

Therefore, balanced incomplete block designs are efficiency-balanced with all canonical efficiency factors equal to E; see (2.11). The designs are clearly optimal under all the optimality criteria conditions given in Section 2.4.

Since the contrast vector $\mathbf{c}$ is an eigenvector of $\mathbf{A}$ with eigenvalue $\lambda v/k$ it follows from the reduced normal equations of (1.11) that $\mathbf{c}'\boldsymbol{\tau}$

is estimated by

$$\mathbf{c}'\hat{\boldsymbol{\tau}} = (k/\lambda v)\mathbf{c}'\mathbf{q} \tag{3.7}$$

with

$$\operatorname{var}(\mathbf{c}'\hat{\boldsymbol{\tau}}) = (k/\lambda v)\mathbf{c}'\mathbf{c}\sigma^2 \tag{3.8}$$

Further, the basic contrasts of the designs are given by any set of $(v-1)$ orthogonal contrasts. As with complete block designs, this means that the $v-1$ degrees of freedom for the treatment sum of squares, adjusted for blocks, can be partitioned into $v-1$ components each with a single degree of freedom by using any set of $v-1$ orthogonal contrasts. Thus, the adjusted treatment sum of squares can be written as

$$\hat{\boldsymbol{\tau}}'\mathbf{q} = (k/\lambda v)\mathbf{q}'\mathbf{q} = \sum_{i=1}^{v-1} (k/\lambda v)(\mathbf{c}_i'\mathbf{q})^2/(\mathbf{c}_i'\mathbf{c}_i) \tag{3.9}$$

for any set of contrast vectors $\mathbf{c}_i$ satisfying $\mathbf{c}_i'\mathbf{c}_j = 0$ $(i, j = 1, 2, \ldots, v-1;$ $i \neq j)$.

In the previous section it was stated that the complement of a balanced incomplete block design, constructed using orthogonal squares, is also a balanced incomplete block design. This result can easily be shown to be true for any balanced incomplete block design. Let $\mathbf{N}$ and $\mathbf{N}_c$ be the incidence matrices of the design and its complement respectively. Then, by definition, $\mathbf{N}_c = \mathbf{J} - \mathbf{N}$. Hence

$$\mathbf{N}_c\mathbf{N}_c' = \mathbf{N}\mathbf{N}' + (b - 2r)\mathbf{J}$$

which is the concurrence matrix of a balanced incomplete block design in which each pair of treatments occurs together in $(\lambda + b - 2r)$ blocks.

3.4 Lattice designs

3.4.1 Resolvability

Consider again the balanced incomplete block design given in Example 3.2 for 9 treatments in 12 blocks of 3 constructed from the complete set of orthogonal 3×3 Latin squares. The design is

(0 1 2)	(0 3 6)	(0 4 8)	(0 5 7)
(3 4 5)	(1 4 7)	(1 5 6)	(1 3 8)
(6 7 8)	(2 5 8)	(2 3 7)	(2 4 6)

The design has been set out in four groups of blocks such that each treatment is replicated exactly once in each group. Each group, therefore, constitutes a replicate. Such an arrangement is an example of a *resolvable* incomplete block design.

Resolvable designs are important in practice since it is often useful to be able to perform an experiment a replicate at a time. In an industrial experiment, for instance, it may not always be possible to carry out all trials in a single session. Use of resolvable designs permits trials to be carried out in stages, with one or more complete replications dealt with at each stage. If the experiment has to be discontinued at any time then all treatments will have occurred equally often. Accuracy will also be increased if the experimental material can be arranged so that replicates are relatively homogeneous. In an agricultural experiment, for example, the land may be divided into a number of large areas corresponding to the replication groups and then each area subdivided into blocks. Resolvable designs are extensively used in variety trials in the United Kingdom; see Patterson and Silvey (1980).

3.4.2 Square lattice designs

Square lattice designs are resolvable designs for $v = s^2$ treatments and $k = s$ plots per block. They are constructed as follows. The s^2 treatments are set out in an $s \times s$ array. In the first replicate, rows of the array correspond to blocks, i.e. all treatments in the jth row of the array are placed in the jth block ($j = 1, 2, \ldots, s$). In the second replicate, blocks correspond to columns of the array. For a third replicate, a Latin square is superimposed on the array and all treatments corresponding to the same letter in the Latin square are placed in the same block. For certain values of s, a fourth replicate can be constructed in a similar way by superimposing onto the array a Latin square orthogonal to the first. Further replicates can be obtained by using Latin squares orthogonal to all previous Latin squares, if such squares exist. The method of construction follows that used to construct the balanced incomplete block designs for $v = n^2$ in Section 3.3.1.

Square lattice designs for $r = 2$ and $r = 3$ replicates are known as *simple lattice* and *triple lattice* designs respectively, and can be constructed for all values of s. For $s = 6$ only simple and triple lattices can be obtained, as no orthogonal Latin squares exist. Lattice designs based on the complete set of mutually orthogonal Latin squares will have $r = s + 1$ replicates and are known as *balanced lattice* designs; they are, of course, also balanced incomplete block designs. Such designs exist for any $s = p^t$, where p is a prime number.

The design for 9 treatments in 12 blocks of 3 given in Example 3.2 is a balanced lattice design for $s = 3$ and $r = 4$. Simple and triple lattices, for $s = 3$, are given by taking the first two and three replicates respectively.

The properties of a square lattice design are most easily derived from its dual design, i.e. from the block concurrence matrix $\mathbf{N'N}$ of the square lattice design. Now the (ij)th element of $\mathbf{N'N}$ will be equal to the number of treatments common to both the ith and jth blocks of the square lattice. For two blocks within the same replicate this element is zero, while for two blocks from different replicates it is 1. Further, the diagonal elements of $\mathbf{N'N}$ are all equal to s. For example, for $s = 3$ the $\mathbf{N'N}$ matrices for square lattices with $r = 2$, 3, 4 are respectively

$$\begin{pmatrix} 3\mathbf{I} & \mathbf{J} \\ \mathbf{J} & 3\mathbf{I} \end{pmatrix}, \quad \begin{pmatrix} 3\mathbf{I} & \mathbf{J} & \mathbf{J} \\ \mathbf{J} & 3\mathbf{I} & \mathbf{J} \\ \mathbf{J} & \mathbf{J} & 3\mathbf{I} \end{pmatrix}, \quad \begin{pmatrix} 3\mathbf{I} & \mathbf{J} & \mathbf{J} & \mathbf{J} \\ \mathbf{J} & 3\mathbf{I} & \mathbf{J} & \mathbf{J} \\ \mathbf{J} & \mathbf{J} & 3\mathbf{I} & \mathbf{J} \\ \mathbf{J} & \mathbf{J} & \mathbf{J} & 3\mathbf{I} \end{pmatrix}$$

In general, for a square lattice design, the matrix $\mathbf{N'N}$ can be partitioned so that all the diagonal matrices are $k\mathbf{I}_s$ and all off-diagonal matrices are $\mathbf{J}_s$, i.e.

$$\mathbf{N'N} = k\mathbf{I}_r \otimes \mathbf{I}_s + (\mathbf{J}_r - \mathbf{I}_r) \otimes \mathbf{J}_s \tag{3.10}$$

where $\otimes$ denotes the Kronecker product (see Section A.6 of the Appendix).

Hence, the dual of a square lattice design has rs treatments which can be divided into r groups of s treatments in such a way that treatments from different groups occur together in a single block while those from the same group do not occur together at all. The dual design is, in fact, a group-divisible PBIB/2 design. Now the vector $\mathbf{p} = \mathbf{p}_1 \otimes \mathbf{p}_2$, where $\mathbf{p}_1$ and $\mathbf{p}_2$ are $r \times 1$ and $s \times 1$ vectors respectively, is an eigenvector of $\mathbf{N'N}$ if $\mathbf{p}_i$ is a contrast vector or if $\mathbf{p}_i = \mathbf{1}$ $(i = 1, 2)$. A full set of eigenvectors can be obtained by taking both $\mathbf{p}_1$ and $\mathbf{p}_2$ to be contrast vectors or equal to $\mathbf{1}$, or by taking one to be a contrast vector and the other equal to $\mathbf{1}$. The canonical efficiency factors of the dual design can then be shown to be $(r-1)/r$ and 1 with multiplicities $r(s-1)$ and $r-1$ respectively. It then follows, using (2.26) and (2.27) and after a little simplification, that the harmonic mean efficiency factor E of a square lattice design with $v=s^2$, $b=rs$ and $k=s$ is

$$E=\frac{(s+1)(r-1)}{r(s+2)-(s+1)} \tag{3.11}$$

For simple, triple and balanced lattices E is respectively

$$\frac{s+1}{s+3},\quad \frac{2s+2}{2s+5},\quad \frac{s}{s+1}$$

For balanced lattices ($r=s+1$) all canonical efficiency factors are equal to $(r-1)/r=s/(s+1)$. For other lattices ($r \leqslant s$) these factors are $(r-1)/r$ and 1 with multiplicities $r(s-1)$ and $v-b+r-1$ respectively.

Finally, a straightforward generalized inverse $\mathbf{\Omega}$ of the information matrix $\mathbf{A}$ of a square lattice design can be obtained using (2.30). From (2.29) and (3.10),

$$\mathbf{A}_d = s\mathbf{I}-(1/r)\mathbf{I}_r\otimes(s\mathbf{I}_s-\mathbf{J}_s)-(1/r)\mathbf{J}_r\otimes\mathbf{J}_s$$

Following (1.37), a generalized inverse of $\mathbf{A}_d$ is given by $\mathbf{\Omega}_d$ where

$$\mathbf{\Omega}_d^{-1}=\mathbf{A}_d+(1/r)\mathbf{J}_r\otimes\mathbf{J}_s=(1/r)\mathbf{I}_r\otimes[s(r-1)\mathbf{I}_s+\mathbf{J}_s]$$

Hence,

$$\mathbf{\Omega}_d=\frac{r}{s(r-1)}\mathbf{I}_r\otimes\left(\mathbf{I}_s-\frac{1}{rs}\mathbf{J}_s\right)$$

Now, since $\mathbf{N}(\mathbf{I}_r\otimes\mathbf{J}_s)=\mathbf{J}_{rs}$ as each $s\times s$ submatrix in $\mathbf{N}$ contains a single unit element in each row and zeros elsewhere,

$$\mathbf{N\Omega}_d\mathbf{N}'=\frac{r}{s(r-1)}\left(\mathbf{NN}'-\frac{1}{s}\mathbf{J}_s\right)$$

From (2.30), and as $\mathbf{AJ}=\mathbf{0}$, it follows that

$$\mathbf{\Omega}=\frac{1}{r}\left[\mathbf{I}+\frac{1}{s(r-1)}\mathbf{NN}'\right] \tag{3.12}$$

The full intra-block analysis for any square lattice design is, consequently, relatively straightforward. No matrix inversions are necessary.

3.4.3 *Rectangular lattice designs*

Rectangular lattice designs are resolvable designs for $v = s(s-1)$ treatments and $k = s - 1$ plots per block. The treatments are initially set out in an $s \times s$ array with the leading diagonal left blank. The blocks of the first two replicates correspond to the rows and columns respectively of the array. Further replicates are obtained by superimposing orthogonal Latin squares onto the array, with the constraint that the leading diagonal of each Latin square contains every treatment in the same order. If there exist $t-1$ mutually orthogonal Latin squares, then $t-2$ of these squares will satisfy this additional constraint. Blocks in each replicate are obtained, as for square lattice designs, by collecting together all treatments corresponding to the same letter in the particular Latin square. Designs with $r = 2$ and $r = 3$ are known as *simple* and *triple rectangular lattices* respectively.

Example 3.4
For $s = 4$, the 12 treatments are set out in a 4×4 array as follows:

.	0	1	2
3	.	4	5
6	7	.	8
9	10	11	.

giving two replicates:

(0 1 2)	(3 6 9)
(3 4 5)	(0 7 10)
(6 7 8)	(1 4 11)
(9 10 11)	(2 5 8)

Superimposing the following orthogonal Latin squares onto the array leads to two further replicates:

A	C	D	B		A	D	B	C
D	B	A	C		C	B	D	A
B	D	C	A		D	A	C	B
C	A	B	D		B	C	A	D

The two replicates are:

(4 8 10)	(5 7 11)
(2 6 11)	(1 8 9)
(0 5 9)	(2 3 10)
(1 3 7)	(0 4 6)

Properties of rectangular lattices can again be obtained most easily be considering the dual designs. The block concurrence matrix $\mathbf{N'N}$ now partitions into r^2, $s \times s$ submatrices with all diagonal submatrices equal to $k\mathbf{I}_s$ and all off-diagonal submatrices equal to $\mathbf{J}_s - \mathbf{I}_s$,

$$\mathbf{N'N} = k\mathbf{I}_r \otimes \mathbf{I}_s + (\mathbf{J}_r - \mathbf{I}_r) \otimes (\mathbf{J}_s - \mathbf{I}_s) \tag{3.13}$$

The canonical efficiency factors of the dual of a rectangular lattice can be shown to be $(rk - s)/rk$, $s(r-1)/rk$ and 1 with multiplicities $k(r-1)$, k and $r-1$ respectively.

Hence, using results in Section 2.8, the canonical efficiency factors of a rectangular lattice are, for $r = s$, $(k-1)/k$ and 1 with multiplicities k^2 and $k-1$ respectively and, for $r < s$, $(rk - s)/rk$, $s(r-1)/rk$ and 1 with multiplicities $k(r-1)$, k and $v - b + r - 1$ respectively.

Rectangular lattice designs were first developed by Harshbarger (1949).

3.4.4 Upper bounds for the efficiency factor of resolvable designs

Lattice designs are only available for a limited number of parameter combinations. Some of the PBIB/2 designs listed in Clatworthy (1973) are also resolvable but usually they are either lattices or less efficient alternatives. Designs based on cyclical methods of construction have been proposed by David (1967) and by Patterson and Williams (1976a) and these will be discussed in the next chapter.

Again, in order to compare alternative resolvable designs and to provide measures against which to judge the best available designs, it will be useful to have upper bounds for the efficiency factor E. Such bounds can be obtained most easily by finding bounds for the dual design and then substituting into (2.27).

In general, the block concurrence matrix $\mathbf{N'N}$ for a resolvable design with $v = ks$ treatments in rs blocks of k plots per block can be partitioned into $s \times s$ submatrices $\mathbf{B}_{ij}$, say $(i, j = 1, 2, \ldots, r)$. The

(ij)th element of $\mathbf{N}'\mathbf{N}$ gives the number of treatments common to both the ith and jth blocks of the design. Two different blocks in the same replicate will have no treatments in common so that $\mathbf{B}_{ii} = k\mathbf{I}_s$ $(i = 1, 2, \ldots, r)$. Consider the s blocks in the ith replicate and one block from the jth replicate $(i \neq j)$. Each of the k treatments from this jth replicate block must occur once, and only once, in one of the blocks in the ith replicate. Hence, the sum of each row of $\mathbf{B}_{ij}$ is equal to k, i.e. $\mathbf{B}_{ij}\mathbf{1} = k\mathbf{1}$ $(i, j = 1, 2, \ldots, r; i \neq j)$. Note that for square lattices $\mathbf{B}_{ij} = \mathbf{J}_s$ from (3.10) and for rectangular lattices $\mathbf{B}_{ij} = \mathbf{J}_s - \mathbf{I}_s$ from (3.13), for all $i \neq j$.

It follows that $(r-1)$ eigenvalues of $\mathbf{N}'\mathbf{N}$ are equal to zero. These correspond to the $(r-1)$ linearly independent eigenvectors representing replication contrasts. One such eigenvector, for instance is given by $(\mathbf{1}'_s, -\mathbf{1}'_s, \mathbf{0}'_s \cdots \mathbf{0}'_s)'$. Hence, the canonical efficiency factors of the dual of a resolvable design are

$$e_1 = e_2 = \cdots = e_{r-1} = 1, \quad e_r, \ldots, e_{b-1}.$$

Their sum, given by the trace of $\mathbf{I} - (1/rk)\mathbf{N}'\mathbf{N}$, is $s(r-1)$ so that

$$\sum_{i=r}^{b-1} e_i = (s-1)(r-1). \tag{3.14}$$

An upper bound for a resolvable design with $v \geqslant b$ can be obtained by replacing the $r(s-1)$ eigenvalues $e_r, \ldots, e_{b-1}$ by their arithmetic mean $(r-1)/r$. Using (2.27), and simplifying, gives the upper bound

$$U_{0r} = \frac{(v-1)(r-1)}{(v-1)(r-1) + (b-r)} \tag{3.15}$$

analogous to U_0 in (2.15) for a general block design. This result was given by Patterson and Williams (1976b). Note that this bound is attained by square lattice designs.

A tighter bound for resolvable designs with $v \geqslant b$, analogous to (2.17), has been given by Jarrett (1977). Williams and Patterson (1977) derive bounds which impose the further constraint that the elements of $\mathbf{B}_{ij}$ $(i \neq j)$ do not differ by more than one.

3.5 Group-divisible designs

For group-divisible designs the $v = mn$ treatments are divided into m groups of n treatments each. The designs are such that all pairs of treatments belonging to the same group occur together in λ_1 (say) blocks, while pairs of treatments from different groups occur together in λ_2 (say) blocks. Treatments in the same group are said to be *first associates* and those from different groups are said to be *second associates.* The arrangement of treatments into groups is called the *association scheme* of the design. It will be assumed that $\lambda_1 \neq \lambda_2$, since if $\lambda_1 = \lambda_2$ the design will be a balanced incomplete block design. The parameters $v, k, r, b, m, n, \lambda_1$ and λ_2 of a group-divisible design must satisfy, in addition to $v = mn$ and (3.3),

$$r(k-1) = (n-1)\lambda_1 + n(m-1)\lambda_2 \qquad (3.16)$$

This is a generalization of (3.4) for balanced incomplete block designs. Now a given treatment will occur with each of its $(n-1)$ first associates in λ_1 blocks and with each of its $n(m-1)$ second associates in λ_2 blocks; hence (3.16) is established. A group-divisible design cannot exist unless both (3.3) and (3.16) are satisfied, although, as was the case with balanced incomplete block designs, such conditions are not sufficient for the existence of a group-divisible design

Example 3.5
Consider the following group-divisible design with $v = 6, r = 2, k = 4, b = 3, m = 3, n = 2$:

$$\begin{matrix} (0 & 1 & 2 & 3) \\ (0 & 1 & 4 & 5) \\ (2 & 3 & 4 & 5) \end{matrix}$$

The association scheme is:

Group	*Treatment*	
1	0	1
2	2	3
3	4	5

The design has $\lambda_1 = 2$ and $\lambda_2 = 1$. For instance, treatments 0 and 1 are first associates and occur together in the first two blocks, while treatments 0 and 2 are second associates and only occur together in the first block.

3.5.1 *Construction of group-divisible designs*

Group-divisible designs have been classified into three subtypes, namely

(i) *Singular*, with $r = \lambda_1$
(ii) *Semi-regular*, with $r > \lambda_1$ and $rk = v\lambda_2$
(iii) *Regular*, with $r > \lambda_1$ and $rk > v\lambda_2$

Bose and Connor (1952) showed that all singular group-divisible designs can be derived from balanced incomplete block designs. Let the labels $1, 2, \ldots, m$ be set out in a balanced incomplete block design with parameters $v^* = m, k^*, r^*, b^*$ and λ^*. Suppose that throughout this design label i is replaced by all n treatments in the ith group of the association scheme ($i = 1, \ldots, m$). The resulting design will be a singular group-divisible design with $v = mn$, $k = nk^*$, $r = r^*$, $b = b^*$, $\lambda_1 = r$ and $\lambda_2 = \lambda^*$. The design given in Example 3.5 is a singular group-divisible design based on the unreduced balanced incomplete block design (1 2), (1 3), (2 3). Label 1 is replaced by treatments 0, 1 of group 1, labels 2 and 3 are replaced by 2, 3 and 4, 5 respectively.

Use is also made of balanced incomplete block designs to construct many of the semi-regular and regular group-divisible designs. In a balanced incomplete block design with $\lambda = 1$, if all the blocks containing one particular treatment are omitted then the resulting design is a group-divisible design. If the balanced incomplete block design is symmetric ($v = b$) then the resulting group-divisible design is semi-regular; otherwise it is regular. If the balanced incomplete block design has parameters v^*, k^*, r^*, b^* and $\lambda = 1$, then the group-divisible design has parameters

$$v = v^* - 1, r = r^* - 1, k = k^*, b = b^* - r^*, m = r^*, n = k^* - 1,$$
$$\lambda_1 = 0, \lambda_2 = 1$$

Example 3.6
Omitting all blocks containing treatment 12 from the design with 13 treatments in Example 3.3 gives

(0 1 2 9)	(0 3 6 10)	(0 4 8 11)
(3 4 5 9)	(1 4 7 10)	(1 5 6 11)
(6 7 8 9)	(2 5 8 10)	(2 3 7 11)

The association scheme of this semi-regular group divisible design is:

Group	Treatment		
1	0	5	7
2	1	3	8
3	2	4	6
4	9	10	11

From any semi-regular group-divisible design, other semi-regular designs can be constructed by omitting all treatments from one or more groups of treatments. In the above design, for instance, omitting the three treatments 9, 10 and 11 from group 4 gives the following design for 9 treatments in 9 blocks of 3:

(0 1 2)	(0 3 6)	(0 4 8)
(3 4 5)	(1 4 7)	(1 5 6)
(6 7 8)	(2 5 8)	(2 3 7)

This design is a semi-regular group-divisible design with $\lambda_1 = 0$ and $\lambda_2 = 1$. It is also a triple lattice.

As already stated in Section 3.3.1, the dual of a square lattice design for s^2 treatments in s plots per block and r^* replicates is a semi-regular design, with $v = r^*s, k = r^*, r = s, b = s^2, m = r^*, n = s, \lambda_1 = 0, \lambda_2 = 1$. The dual of any resolvable balanced incomplete block design, including of course balanced lattices, is also a semi-regular design.

Numerous other methods have been used to construct semi-regular and regular group-divisible designs; see, for instance, John (1971), Raghavarao (1971), Clatworthy (1973), Freeman (1976b) and John and Turner (1977).

3.5.2 Properties of group-divisible designs

It will be assumed that the association scheme of the group-divisible designs is such that the ith group consists of the n treatments

$$(i-1)n+1, (i-1)n+2, \ldots, in$$

for $i = 1, \ldots, m$. Then the concurrence matrix $\mathbf{NN'}$ can be partitioned into m^2, $n \times n$ submatrices. The diagonal submatrices have all diagonal elements equal to r and all off-diagonal elements equal to λ_1, while all elements of the off-diagonal submatrices are equal to λ_2, i.e.

$$\mathbf{NN}' = \mathbf{I}_m \otimes [(r - \lambda_1)\mathbf{I}_n + \lambda_1 \mathbf{J}_n] + (\mathbf{J}_m - \mathbf{I}_m) \otimes \lambda_2 \mathbf{J}_m.$$

The information matrix $\mathbf{A}$ is then

$$\mathbf{A} = (1/k)\mathbf{I}_m \otimes [(rk - r + \lambda_1)\mathbf{I}_n - (\lambda_1 - \lambda_2)\mathbf{J}_n] - (\lambda_2/k)\mathbf{J}_{mn} \qquad (3.17)$$

The eigenvectors of $\mathbf{A}$, the basic contrasts of the design, are of the form

$$\mathbf{p}_1 = \mathbf{a}_m \otimes \mathbf{c}_n \qquad (3.18)$$

or

$$\mathbf{p}_2 = \mathbf{c}_m \otimes \mathbf{1} \qquad (3.19)$$

where $\mathbf{a}_m$ is any $m \times 1$ vector and $\mathbf{c}_m$ and $\mathbf{c}_n$ are contrast vectors of length m and n respectively. The canonical efficiency factors of a group-divisible design are, therefore, $e_1 = (rk - r + \lambda_1)/rk$ with multiplicity $m(n-1)$, and $e_2 = v\lambda_2/rk$ with multiplicity $m-1$, corresponding to the eigenvectors $\mathbf{p}_1$ and $\mathbf{p}_2$ respectively. For singular designs $e_1 = 1$, whereas for semi-regular designs $e_2 = 1$.

The harmonic mean of the canonical efficiency factors is, after some simplification,

$$E = m(v-1)\lambda_2 e_1/[m(v-1)\lambda_2 + (m-1)(\lambda_1 - \lambda_2)] \qquad (3.20)$$

If $\mathbf{a}_m$ is a column of the identity matrix and if $\mathbf{c}_n$ is any pairwise contrast vector, then the difference between any two treatments from the same group is a basic contrast of the group-divisible design, i.e. $\tau_i - \tau_j$ is a basic contrast if treatments i and j are first associates. There are $m(n-1)$ such pairwise contrasts, each having efficiency factor e_1.

The other basic contrasts of the design involve the sums of treatment effects over all treatments belonging to the same group. Since there are m such sums, there are $m-1$ linearly independent normalized contrasts between these sums, each having efficiency factor e_2.

Pairwise contrasts involving two treatments from different groups can be expressed as linear combinations of $\mathbf{p}_1$ and $\mathbf{p}_2$. For instance, let

$$\mathbf{p}_1' = (1 \;\; -1 \;\; 0 \;\; \cdots \;\; 0) \otimes (n-1 \;\; -1 \;\; \cdots \;\; -1)$$
$$\mathbf{p}_2' = (1 \;\; -1 \;\; 0 \;\; \cdots \;\; 0) \otimes (1 \;\; 1 \;\; \cdots \;\; 1)$$

then $\mathbf{c} = (\mathbf{p}_1 + \mathbf{p}_2)/n$ represents a pairwise contrast between the first treatments in the first and second groups respectively. The efficiency factor E_2 for the difference between two treatments which are second associates can then be shown to be

$$E_2 = \frac{ne_1e_2}{e_1 + (n-1)e_2} = \frac{v\lambda_2 e_1}{v\lambda_2 + (\lambda_1 - \lambda_2)}$$

Note that when $\lambda_1 = \lambda_2$, $e_1 = e_2 = E_2$, i.e. the design is a balanced incomplete block design. Also, that E_2, and hence E, are maximized when $|\lambda_1 - \lambda_2|$ is minimized, thereby supporting the conjecture made in Section 2.6 that the A-optimal (and D- and E-optimal) designs are in the class of (M, S)-optimal designs.

Group-divisible designs are particularly useful in factorial experiments, where the treatments are combinations of a number of different treatment factors. As will be shown in Section 7.3, with a suitable choice of vectors $\mathbf{a}_m$, $\mathbf{c}_m$ and $\mathbf{c}_n$, the basic contrasts of the designs will correspond to the treatment contrasts of interest in a factorial experiment.

CHAPTER 4

Cyclic designs

4.1 Introduction

Many of the designs discussed in the previous chapter can be obtained very simply by a method of cyclic substitution. For example, a balanced incomplete block design for seven treatments in blocks of three is given by the seven blocks

(0 1 3)
(1 2 4)
(2 3 5)
(3 4 6)
(4 5 0)
(5 6 1)
(6 0 2)

Each treatment is replicated three times and every pair of treatments occurs together in a single block. The cyclic method of construction is such that a block is obtained by adding one to each element in the previous block and reducing modulo 7 when necessary. The full design can, therefore, be generated from one of the blocks, which is called the *initial block*. For convenience, the initial block will be taken to be the block of lowest numerical value. Hence, the above design is constructed by a cyclic development of the initial block (0 1 3).

For a second example, the cyclic development of the initial block (0 1 3) with six treatments in blocks of three will give a group-divisible design in six blocks. The six treatments fall into the three groups (0, 3), (1, 4) and (2, 5), with treatments from the same group occurring together in two blocks and those from different groups in a single block.

Another group-divisible design with eight treatments in blocks of three is obtained by cyclic development of the initial blocks (0 1 2), (0 1 4) and (0 2 5). In this case, each initial block generates a set of eight blocks, so that the full design has 24 blocks with each treatment replicated nine times. The eight treatments fall into the two groups (0, 2, 4, 6) and (1, 3, 5, 7), with treatments from the same and different groups occurring together in two and three blocks respectively.

These examples show how a method of cyclic substitution can be used to construct designs. Many lattice, balanced incomplete block and group-divisible designs are cyclic, but many are not. Cyclic designs, together with variants based on the cyclical method of construction, will be considered in detail in this chapter.

4.2 Cyclic designs

Cyclic designs are incomplete block designs consisting in the simplest case of a set of blocks obtained by cyclic development of an initial block. More generally, they consist of combinations of such sets and will be said to be of *size* (v, k, r), where v is the number of treatments, k the block size and r the number of replications. All cyclic designs belong to the class of partially balanced (PBIB) designs of Bose and Nair (1939) but may have up to $v/2$ associate classes. Cyclic designs exist for many parameter combinations for which no balanced incomplete block or PBIB/2 design is available, although the reverse is also true. As the examples in the previous section illustrate, many balanced incomplete block and PBIB/2 designs may be set out as cyclic designs (see David and Wolock, 1965; John 1969; Clatworthy, 1973).

For given v and k the $\binom{v}{k}$ distinct blocks can be set out in a number of cyclic sets. For example, for $v = 7$ and $k = 3$ the 35 distinct blocks can be set out in five cyclic sets each of $b = 7$ blocks as follows:

012	123	234	345	456	560	601
013	124	235	346	450	561	602
014	125	236	340	451	562	603
015	126	230	341	452	563	604
024	135	246	350	461	502	613

If these sets are used singly or in combination, cyclic designs of size $(7, 3, r)$ can be constructed where r is any multiple of 3.

The five sets above are *full* sets consisting of $b = v$ blocks. If v and k are not relatively prime then *partial* sets consisting of v/d blocks arise, where d is any common divisor of v and k. For example, with $v=8$ and $k=4$ the 70 distinct blocks can be set out in 8 full sets of 8 blocks; one half-set of four blocks given by

0145 1256 2367 3470;

and one quarter-set of two blocks given by

0246 1357

Hence, in this case, the full and partial sets can be used to give cyclic designs for any value of $r \geqslant 1$, although the design for $r=1$ is clearly disconnected.

In general, a partial set is constructed by taking as the initial block a subgroup of the treatments of order d together with $(k/d)-1$ of its cosets. With $p=v/d$, the treatments

$$0, p, 2p, \ldots, (d-1)p$$

constitute a subgroup of the v treatment labels closed to addition modulo v. The treatments

$$j, j+p, j+2p, \ldots, j+(d-1)p \qquad (j=1,2,\ldots,p-1)$$

form the cosets of this subgroup. For example, with $v=8$ and $k=4$, the subgroup of size $d=2$ is 0, 4 with cosets 1, 5; 2, 6; and 3, 7. Note that if the initial block contains the subgroup 0, 4 and coset 2, 6 the partial set will have 2, not 4, distinct blocks, as 0, 2, 4, 6 is itself a subgroup of order 4.

Full and partial sets can be used singly or in combination, thereby giving a flexible class of designs. A cyclic design of size $(v, k, r=ik)$ exists for all positive integers v, k and i. If v and k have a common divisor d then a partial set of size $(v, k, r=k/d)$ exists corresponding to each d. These partial sets may also be combined with the full sets or with other partial sets to form further designs.

Apart from their flexibility, cyclic designs have other advantages which make them attractive. No plan of the experimental layout is needed since the initial blocks are sufficient. This also leads to ease in experimentation. Once treatment labels have been randomly assigned, constant reference to an experimental plan becomes unnecessary. In view of their method of generation, cyclic designs of size $(v, k, r=ik)$ provide automatic elimination of heterogeneity in two directions. In the sets for $v=7$ and $k=3$, for instance, it can be seen that each treatment appears once in each position within a block. Consequently, position effects can readily be eliminated in the analysis if desired. This point is considered more fully in Chapter 5 on row–column designs. Finally, since the information matrix of a cyclic design is circulant an explicit expression for the canonical efficiency factors can be obtained.

For given v, k and r the most efficient cyclic designs can be determined by simply evaluating and comparing the efficiency factors for each design in the class. The amount of computation necessary can be drastically reduced if use is made of the fact that many cyclic sets and designs are equivalent to each other, i.e. can be derived from each other by a relabelling or permutation of the treatments.

As an example, consider again the five sets of size (7, 3, 3) given above. Suppose that treatment i in the set with initial block (0, 1, 2) is replaced by treatment $3i \pmod v$ for all values of i. The resulting design will be the cyclic set with initial block (0 1 4). Set (0 1 2) is, therefore, equivalent to set (0 1 4). Repeated application of this permutation will establish that the five sets fall into two equivalence groups, with (0 1 2), (0 1 4) and (0 2 4) in one and (0 1 3) and (0 1 5) in the other.

Full details of this permutation theory is given in John, Wolock and David (1972). Such a theory is an invaluable aid in the search for good designs, since it leads to a considerable reduction in the number of designs that have to be considered.

4.3 Efficiency factors of cyclic designs

A feature arising from the method of constructing cyclic sets is that the k successive differences between treatment labels in any block are invariant throughout the set. This means that, for instance, if treatments 0 and 1 occur together in, say, λ_1 blocks then so also will treatments 1 and 2, 2 and 3, and so on. Consequently the concurrence matrix $\mathbf{NN}'$, the information matrix $\mathbf{A}$ of (1.12) and its generalized inverse $\boldsymbol{\Omega}$ given by (1.17) or (1.37) are circulant matrices.

Let treatments 0 and i occur together in λ_i blocks ($i = 1, 2, \ldots, v-1$) and let $\lambda_0 = r$. Then the concurrence matrix $\mathbf{NN}'$ will take the form

$$\mathbf{NN}' = \begin{pmatrix} \lambda_0 & \lambda_1 & \lambda_2 & \cdots & \lambda_{v-1} \\ \lambda_{v-1} & \lambda_0 & \lambda_1 & \cdots & \lambda_{v-2} \\ \lambda_{v-2} & \lambda_{v-1} & \lambda_0 & \cdots & \lambda_{v-3} \\ \vdots & & & & \\ \lambda_1 & \lambda_2 & \lambda_3 & \cdots & \lambda_0 \end{pmatrix}$$

where $\lambda_i = \lambda_{v-i}$ $(i > 0)$ since $\mathbf{NN'}$ is symmetric. This circulant matrix can be specified by the elements in the first row, since the other rows are obtained from the first row by a cyclical rotation.

Let $\boldsymbol{\Gamma}_h$ be the $v \times v$ circulant matrix whose first row has 1 in the $(h+1)$th column and zero elsewhere. Then $\mathbf{NN'}$ can be written more concisely as

$$\mathbf{NN'} = \sum_{h=0}^{v-1} \lambda_h \boldsymbol{\Gamma}_h.$$

Thus, the information matrix $\mathbf{A}$ given in (1.36) can be written as

$$\mathbf{A} = \sum_{h=0}^{v-1} a_h \boldsymbol{\Gamma}_h \tag{4.1}$$

where

$$a_0 = r(k-1)/k, \qquad a_h = -\lambda_h/k \qquad (h = 1, 2, \ldots, v-1)$$

Since $\mathbf{A}$ is a symmetric matrix then, using (A.32) of the Appendix, the canonical efficiency factors of a connected cyclic design are given by

$$e_u = (k-1)/k - (1/rk) \sum_{h=1}^{v-1} \lambda_h \cos(2\pi hu/v) \qquad (u = 1, 2, \ldots, v-1) \tag{4.2}$$

From (A.35), the generalized inverse of $\mathbf{A}$ given in (1.17) can be written as

$$\boldsymbol{\Omega} = \sum_{h=0}^{v-1} \theta_h \boldsymbol{\Gamma}_h$$

where

$$\theta_h = (1/rv) \sum_{u=1}^{v-1} e_u^{-1} \cos(2\pi hu/v) \qquad (h = 0, 1, \ldots, v-1) \tag{4.3}$$

Hence, the efficiency factor E_{ij} for the pairwise comparison $\tau_i - \tau_j$ $(i < j)$ is, using (2.1), (4.3) and the fact that $\boldsymbol{\Omega}$ is circulant,

$$E_{ij} = [r(\theta_0 - \theta_l)]^{-1} \qquad l = j - i. \tag{4.4}$$

Note also that

$$E = \frac{v-1}{rv\theta_0} \tag{4.5}$$

where E is the harmonic mean of the canonical efficiency factors or of the pairwise efficiency factors.

Table 4.1 *Efficiency factors for cyclic designs with* $v = 11$, $k = r = 4$.

Initial block	*E*	e_{min}	e_{max}
(0 1 2 3)	.702	.348	.991
(0 1 2 4)	.788	.535	.948
(0 1 2 5)	.817	.707	.932
(0 1 3 4)	.794	.605	.997

The efficiency factors given in (4.2) and (4.4) can be easily calculated on a desktop microcomputer or even on a programmable calculator. It is unnecessary with cyclic designs to use general, and time-consuming, eigenvalue and inverse subroutines.

Example 4.1
Consider cyclic designs of size (11, 4, 4). The 330 distinct blocks can be set out in 30 cyclic sets. Under a permutation of the treatment labels each of these sets can be shown to be equivalent to one of four sets. The four designs together with average efficiency factor E and the smallest ($e_{\min}$) and largest ($e_{\max}$) canonical efficiency factors are shown in Table 4.1. There are large differences between the efficiency factors of the designs. The design with initial block (0 1 2 5) is clearly the best design, on the basis of the optimality criteria given in Section 2.4.

4.4 Construction of efficient designs

It has already been seen in (4.2) that the off-diagonal elements of the concurrence matrix $\mathbf{NN}'$ are required in order to calculate the canonical efficiency factors. Further, according to the criteria in Section 2.5 for choosing efficient designs, the set of cyclic designs with high efficiency factors are those which minimize the range of these off-diagonal elements. The search for good designs is facilitated by the fact that the values of the $\lambda_i (i > 0)$ can be readily obtained from the initial blocks of the design by a method of differencing.

For the full cyclic set with initial block $(0\ x_1\ x_2 \ldots x_{k-1})$ all possible differences between pairs of treatments are obtained to give the difference set

$$[x_1, x_2, \ldots, x_{k-1}, x_2 - x_1, \ldots, x_{k-1} - x_1, \ldots, x_{k-1} - x_{k-2}].$$

Any difference i greater than m is replaced by $v-i$, where $m=(v-1)/2$ for v odd and $m=v/2$ for v even. If a_i represents the number of differences equal to i then this difference set can be represented more briefly by

$$[1^{a_1}, 2^{a_2}, \ldots, m^{a_m}]. \tag{4.6}$$

Then $\lambda_i = \lambda_{v-i} = a_i$ for all i, except that $\lambda_m = 2a_m$ when v is even. For example, for $v=7$ and $k=4$ the initial block (0 1 3 5) gives the paired difference set [1 3 2 2 3 2] or $[1^1, 2^3, 3^2]$. Hence, $\lambda_1 = \lambda_6 = 1$, $\lambda_2 = \lambda_5 = 3$ and $\lambda_3 = \lambda_4 = 2$. For $v=8$ and $k=4$ the initial block (0 1 4 6) gives [1 4 2 3 3 2] or $[1^1, 2^2, 3^2, 4^1]$, so that $\lambda_1 = \lambda_7 = 1$, $\lambda_2 = \lambda_6 = 2$, $\lambda_3 = \lambda_5 = 2$ and $\lambda_4 = 2$.

For partial sets with v/d blocks, where d is a common divisor of v and k, the same differencing procedure is used except that the λ_i obtained from (4.6) have to be further divided by d. For example, for $v=8$ and $k=4$ a partial set with 4 blocks ($d=2$) is obtained from the initial block (0 1 4 5). Since this gives [1 4 3 3 4 1] or $[1^2, 2^0, 3^2, 4^2]$, then $\lambda_1 = \lambda_7 = 1$, $\lambda_2 = \lambda_6 = 0$, $\lambda_3 = \lambda_5 = 1$ and $\lambda_4 = 2$. For designs with more than one initial block the concurrences are obtained by adding together the λ_i values from the separate initial blocks. A design of size (8, 4, 6) with initial blocks (0 1 4 5) and (0 1 4 6) has, therefore, $\lambda_1 = \lambda_2 = \lambda_6 = \lambda_7 = 2$, $\lambda_3 = \lambda_5 = 3$ and $\lambda_4 = 4$.

A catalogue of efficient cyclic designs has been given by John, Wolock and David (1972). Four tables of designs are listed for the following values of v, k and r:

Table A:	$k=2$,	$r \leqslant 10$,	$6 \leqslant v \leqslant 30$
Table B:	$k \geqslant 3$,	$r \geqslant k \leqslant 10$,	$6 \leqslant v \leqslant 15$
Table C:	$k \geqslant 3$,	$r = k \leqslant 10$,	$16 \leqslant v \leqslant 30$
Table D:	$k \geqslant 3$,	$r < k \leqslant 10$,	$10 \leqslant v \leqslant 60$

For Tables A, B and D all the non-isomorphic designs were enumerated, using the permutation theory referred to in Section 4.2. For a given size the best design, in the sense of maximizing the overall efficiency factor E of (4.5), was listed. In Table C attention was restricted to the class of designs having treatments occurring together in either λ_1 or λ_2 blocks, where $\lambda_2 = \lambda_1 + 1$. The best designs were chosen from this class, although for some parameter combinations there were no designs in the class, thus resulting in a few omissions in the table.

The designs in Table D have $r < k$ and are, therefore, based on partial cyclic sets. In general, such designs can be relatively inefficient. This is because the initial block of a partial set of $p = v/d$ blocks, where d is a common divisor of v and k, must consist of the subgroup $0, p, 2p, \ldots, (d-1)p$ and a number of its cosets, as explained in Section 4.2. Thus, from the difference set (4.6), it follows that $\lambda_p = \lambda_{2p} = \cdots = \lambda_{(d-1)p} = k/d$ which may be considerably different from other λ_i values. When $r < k$ it is usually preferable, therefore, to use the dual of a cyclic design for $r > k$ rather than a design from Table D. Although the dual of a cyclic design may not be cyclic, many can be obtained by cyclical methods of construction as is shown in Section 4.6.

As an alternative to the extensive catalogue referred to above, John (1981) has provided two small compact tables of efficient cyclic designs which can be used whenever $6 \leqslant v \leqslant 30$, $r \leqslant 10$, and r is a multiple of k. In a few cases the designs in these tables are slightly less efficient than those in the catalogue of John, Wolock and David (1972). For $r < k$ the use of the dual of a design with $r > k$ is suggested.

The two tables of designs are reproduced here as Tables 4.2 and 4.3. Table 4.2 provides a list of efficient cyclic designs for $4 \leqslant v \leqslant 15$; designs for $v = 4$ and 5 have been added to those in John (1981). For given k, the first $k - 1$ treatments in the initial block are taken to be the same for all v, with the kth and final treatment depending on the value of v. For example, for $r = k = 4$ the initial block for $v = 7$ is (0 1 3 6) and that for $v = 8$ is (0 1 3 7). Designs for $r > k$ are obtained by adding further sets of v blocks. For instance, the design of size (7, 4, 8) is given by two initial blocks, namely (0 1 3 6) used for $r = k = 4$ and (0 1 4 6). Of the 173 designs listed in Table 4.2, 152 are A-optimal. A-optimality here and elsewhere in this section will mean that a design is A-optimal within the class of cyclic designs; it does not necessarily mean that it is A-optimal over the entire class of designs of that size. Similar remarks apply to the use of E-optimality. For 11 cases, denoted by asterisks in Table 4.2, there are virtually no differences between the values of the average factor E of the designs given in the table and the A-optimal cyclic designs. For the remaining 10 cases alternative designs are given in Table 4.4.

Table 4.3 lists designs for $16 \leqslant v \leqslant 30$. For $k = 2$ and for most cases with $r = k > 2$ a comparison with the designs in the catalogue of John et al. (1972) can be made. For all other cases, the designs listed in Table 4.3 are the results of extensive, but not exhaustive, computer searches for efficient designs. Of the 254 designs in

Table 4.2 *Efficient cyclic designs for* $4 \leqslant v \leqslant 15$, $r \leqslant 10$, $r = ik$ $(i \geqslant 1)$.

k	r	*First* $k-1$ *treatments*	*kth treatment*, $v =$ 4	5	6	7	8	9	10	11	12	13	14	15
	2	0	1	1	1	1	1	1	1	1	1	1	1	1
	4	0	2	2	2	3	3	3	3*	3	3	5	4	4
2	6	0	1	3	3	2	2	2	2	5	5	2	6	2
	8	0	1	4	5	4	4	4	4	2	2	4	3	7*
	10	0	2	1	4	5	5	5	5	4	4	3	5	5
	3	0 1	2	2	3	3	3	3	4	4	4	4	4	4
3	6	0 2	1	3	1	3	7	6	7	7	5	7	7	8
	9	0 1	3	2	3	3	4	5	3	3	6	4*	6	5
4	4	0 1 3	—	2	2	6	7	7	6	7	7	9	7	7
	8	0 1 4	—	2	2	6	7	8	2	6	6*	6	6	6*
	5	0 1 2 4	—	—	5	5	7	7	7	7	7	7	9	10
5	10	{0 2 3 4	—	—	5	5	7	8	9	8				
		{0 2 3 6									7	11	12	10
6	6	0 1 2 3 6	—	—	—	5	5	5	5	10	10*	10	10	10
7	7	0 1 2 3 4 7	—	—	—	—	5	5	9	9	9	9*	9*	10
8	8	0 1 2 3 4 6 8	—	—	—	—	—	5	9	9	9	9*	11	11
9	9	0 1 2 3 4 5 7 9	—	—	—	—	—	—	8	8	8	10	10	10*
10	10	0 1 2 3 4 5 6 9 10	—	—	—	—	—	—	—	7	7	7	12	12*

Reproduced from John (1981) with permission of the Royal Statistical Society

Table 4.3 *Efficient cyclic designs for* $16 \leqslant v \leqslant 30$, $r \leqslant 10$, $r = ik$ $(i \geqslant 1)$.

k	r	First $k-1$ treatments	kth treatment, $v=$ 16	17	18	19	20	21	22	23	24	25	26	27	28	29	30
	2	0	1	1	1	1	1	1	1	1	1	1	1	1	1	1	1
	4	0	4	4	5	5	8	8	5	5	10	10	10	11	8	6	5
2	6	0	7	7	3	8	5	3	8	8	6	3	4	7	5	9	8
	8	0	2	2	7	2	2	6	10	11	3	8	7	3	11	13	12
	10	0	6	5	9	4	9	4	3	2	8	6	2	5	2	2	14
	3	0 1	4	4	4	8	6	5	5	5	5	5	5	5	5	5	6
3	6	0 2	7	8	7	5	9	10	10	10	10	10	11	10	11	12	14
	9	{0 2	8	7	8												
		{0 3					8	9	9	9	9	12	10	14	13	11	10
	4	{0 1 3	7	7	7	8	8	9	9	17	10						
		{0 1 6										9	9	9	9	9	9
4	8	{0 1 5	8	10	10	11	11	12	12	13	13						
		{0 2 12										16	13	16	13	16	19
5	5	0 2 7 8	3	3	11	11	11	11	11	11	11	11	11	11	17	20	18
	10	{0 2 5 6	9	9						13	14					15	
		{0 1 4 14			16	16	16	16	16			19	21	12	26		19

6	6	0	1	2	6	9						15	16			11	11	11	14	15				17	19
		0	2	8	9	12					11			13	13						13	13	13		
7	7	0	1	2	3	8	12				5	5	5	6											
		0	1	2	4	8	11								16	16				16			16		
		0	2	7	8	10	11										17	14	20		21	22		24	23
8	8	0	1	2	3	6	8	11			15	15	15	15	15	15	15	15	15						
		0	1	2	4	8	11·16													21	21	21	21	21	21
9	9	0	1	2	3	6	8	11	15		12						12	19	22	19	18	17	19	19	20
		0	2	3	4	5	8	11	12			15	16	17	18	19									
10	10	0	1	3	4	7	8	9	10	12	14	14	14	14	14	14									
		0	1	2	3	6	8	12	15	19							20	21	11	11	11	11	20	27	11

Reproduced from John (1981) with permission of the Royal Statistical Society

Table 4.3, 157 are A-optimal or the best found in the searches. Most of the other designs have E values which differ by less than 0.003 from the A-optimal or best found designs. For those cases where the differences are greater than 0.003, alternative designs are given in Table 4.4. No design of size (19, 3, 9) is listed in Table 4.3 since a balanced incomplete block design is given in Table 4.4.

In checking on the E-optimality of the designs, John (1981) concludes that if there is little to choose between designs on the A-optimality criterion, there is little difference on the E-optimality criterion. Alternative designs are listed in John (1981) for 15 cases where the smallest canonical efficiency factor is at least 0.04 higher than in the designs in Table 4.2 and 4.3.

Table 4.4 *Alternative designs.*

k	r	v	*Initial blocks*
2	4	12	(0 2)(0 3)
2	4	24	(0 3)(0 4)
2	4	30	(0 3)(0 5)
2	6	9	(0 1)(0 2)(0 4)
2	6	14	(0 1)(0 3)(0 5)
2	6	15	(0 1)(0 4)(0 6)
2	6	18	(0 1)(0 3)(0 8)
2	8	12	(0 1)(0 2)(0 4)(0 5)
2	8	16	(0 1)(0 3)(0 5)(0 7)
2	8	24	(0 2)(0 3)(0 9)(0 10)
2	10	20	(0 1)(0 3)(0 5)(0 7)(0 9)
2	10	22	(0 1)(0 3)(0 5)(0 7)(0 9)
2	10	24	(0 1)(0 3)(0 5)(0 7)(0 9)
3	6	8	(0 1 3)(0 1 5)
3	6	30	(0 1 6)(0 3 13)
3	9	19	(0 1 8)(0 2 5)(0 4 10)
3	9	20	(0 1 5)(0 2 10)(0 3 9)
4	4	14	(0 1 4 6)
6	6	11	(0 1 2 4 5 7)
7	7	15	(0 1 2 4 5 8 10)
8	8	15	(0 1 2 3 7 10 11 13)

Tables of symmetric cyclic designs for $v > 30$ have been given by Lamacraft and Hall (1982). Also included are some designs omitted from the catalogue by John et al. (1972), namely designs whose complement had been given or whose concurrences differ by more than one. For completeness they give a full listing of designs in the range $10 \leqslant v \leqslant 60$, $3 \leqslant r = k \leqslant 10$.

4.5 *n*-Cyclic designs

The cyclic designs considered in the previous sections have been obtained by a cyclic development of one or more initial blocks. Two important features of a cyclic design are that each treatment label is represented by a single number and that the elements of a block are obtained by adding one to the elements in the previous block, reducing modulo v when necessary. Two generalizations will now be considered. In the first, each treatment label is represented by a set of numbers and the cyclic development is carried out on each number in turn. The resulting designs will be called *n-cyclic designs* and will be considered in this section; these designs have also been called GC/n designs. In the second generalization a value greater than one is added to the elements of a block to give the next block. These designs will be called *generalized cyclic designs* and will be considered in Section 4.6.

Two of the balanced incomplete block designs listed in the tables of Fisher and Yates (1963) are given by dicyclic (i.e. 2-cyclic) solutions. One of these designs has 16 treatments in 16 blocks of 6. The 16 treatments are represented by 4 letters in combination with 4 suffices. Four blocks are first obtained by cyclic development of the letters in the initial block ($a_1\ a_2\ a_3\ b_1\ c_4\ d_1$). The full dicyclic design is then given by a cyclic development of the suffices in each of these four blocks.

Alternatively, each treatment can be represented by a pair of digits $a_1\ a_2$, say, where $a_i = 0, 1, 2, 3$ $(i = 1, 2)$. Hence, the 16 treatments are, in order of magnitude

00 01 02 03 10 11 12 13 20 21 22 23 30 31 32 33.

The balanced incomplete block design is obtained by a cyclic development of each digit in turn on the initial block (00 01 02 10 23 30). The 16 blocks are

(00	01	02	10	23	30)	(20	21	22	30	03	10)
(01	02	03	11	20	31)	(21	22	23	31	00	11)
(02	03	00	12	21	32)	(22	23	20	32	01	12)
(03	00	01	13	22	33)	(23	20	21	33	02	13)
(10	11	12	20	33	00)	(30	31	32	00	13	20)
(11	12	13	21	30	01)	(31	32	33	01	10	21)
(12	13	10	22	31	02)	(32	33	30	02	11	22)
(13	10	11	23	32	03)	(33	30	31	03	12	23)

Each treatment label in a dicyclic design is made up of two digits, and as such is said to be a 2-factor design. More generally, an n-cyclic design will have n factors and $v = m_1 m_2 \ldots m_n$ treatments, where the ith factor is said to be at m_i levels ($i = 1, 2, \ldots, n$). A treatment is represented by an n-tuple $a_1 a_2 \ldots a_n$ where $a_i = 0, 1, \ldots, m_i - 1$ ($i = 1, 2, \ldots, n$). Treatments can be written in order of magnitude and identified with the numbers $1, 2, \ldots, v$. For example, with $n = 3$, $m_1 = 2$, $m_2 = 2$ and $m_3 = 3$, the $v = 12$ treatments are

000 001 002 010 011 012 100 101 102 110 111 112

so that, for instance, the third treatment is 002 and the eighth 101.

An n-cyclic set is generated from an initial block consisting of k treatments. The jth block of the set is given by adding the jth treatment to each treatment in the initial block, where addition is defined as

$$a_1 a_2 \ldots a_n + b_1 b_2 \ldots b_n = c_1 c_2 \ldots c_n \tag{4.7}$$

where $c_i = a_i + b_i \bmod m_i$ ($i = 1, 2, \ldots, n$). This method of generating a set is equivalent to a cyclical development of the initial block where each digit in the n-tuple is cycled in turn.

Example 4.2

As a further example, consider the set of $v = 12$, $n = 3$, $m_1 = m_2 = 2$, $m_3 = 3$ and $k = 4$ obtained from the initial block (000 011 101 112). The full set of twelve blocks is

(000	011	101	112)	(001	012	102	110)	(002	010	100	111)
(010	001	111	102)	(011	002	112	100)	(012	000	110	101)
(100	111	001	012)	(101	112	002	010)	(102	110	000	011)
(110	101	011	002)	(111	102	012	000)	(112	100	010	001)

Note that the blocks in any column are obtained by a cyclic development of the first two digits with the second digit cycled first. The blocks in any row are obtained by a development of the third digit.

The two designs given above are examples of full sets with $b = v$ blocks. Partial sets can also be constructed using a method similar to that used in Section 4.2 to obtain partial cyclic sets. The v treatment labels form a group G closed to addition defined by (4.7). Let S be a subgroup of G of order d, where d is a common factor of v and k. The initial block of a partial set of v/d blocks is then given by S together with $(k/d) - 1$ of its cosets. For instance, for $v = 12, n = 3$, $m_1 = m_2 = 2, m_3 = 3$ and $k = 4$, a subgroup of order 2 is given by 000 110 while a subgroup of order 4 is given by 000 010 100 110. Cosets are obtained by adding further treatments to every element in the subgroup. Hence, for example, initial blocks for sets with 6 and 3 blocks are (000 001 110 111) and (000 010 100 110) respectively.

By taking partial or full sets singly or in combination, a large and flexible class of n-cyclic designs is available. Two of the balanced incomplete block designs listed in the tables of Fisher and Yates (1963) are 2-cyclic designs. One is given above. The other has 25 treatments in 50 blocks of 4 and is made up of two 2-cyclic sets with initial blocks (00 01 10 44) and (00 02 20 33), where $m_1 = m_2 = 5$. P.W.M. John (1971) uses n-cyclic designs to construct various group-divisible, Latin square type and cubic partially balanced designs. Freeman (1976b) obtains a number of new group-divisible designs using the n-cyclic method of construction. n-Cyclic designs have also been shown to be particularly useful in the construction of designs for factorial experiments, as will be shown in Chapters 6 and 7.

In the analysis of n-cyclic designs, the concurrence matrix $\mathbf{NN}'$, the information matrix $\mathbf{A}$ of (1.12) and its generalized inverse matrix $\mathbf{\Omega}$ given by (1.17) or (1.37) will all be block circulant matrices. The notation used in Section 4.3 for circulant matrices can be extended to block circulants; see also the Appendix. To fix ideas, consider the concurrence matrix $\mathbf{NN}'$ for the design for $v = b = 12$ and $r = k = 4$ given in Example 4.2, namely:

$$\mathbf{NN}' = \left(\begin{array}{ccc|ccc|ccc|ccc} 4 & 0 & 0 & 0 & 2 & 2 & 0 & 2 & 2 & 2 & 1 & 1 \\ 0 & 4 & 0 & 2 & 0 & 2 & 2 & 0 & 2 & 1 & 2 & 1 \\ 0 & 0 & 4 & 2 & 2 & 0 & 2 & 2 & 0 & 1 & 1 & 2 \\ \hline 0 & 2 & 2 & 4 & 0 & 0 & 2 & 1 & 1 & 0 & 2 & 2 \\ 2 & 0 & 2 & 0 & 4 & 0 & 1 & 2 & 1 & 2 & 0 & 2 \\ 2 & 2 & 0 & 0 & 0 & 4 & 1 & 1 & 2 & 2 & 2 & 0 \\ \hline 0 & 2 & 2 & 2 & 1 & 1 & 4 & 0 & 0 & 0 & 2 & 2 \\ 2 & 0 & 2 & 1 & 2 & 1 & 0 & 4 & 0 & 2 & 0 & 2 \\ 2 & 2 & 0 & 1 & 1 & 2 & 0 & 0 & 4 & 2 & 2 & 0 \\ \hline 2 & 1 & 1 & 0 & 2 & 2 & 0 & 2 & 2 & 4 & 0 & 0 \\ 1 & 2 & 1 & 2 & 0 & 2 & 2 & 0 & 2 & 0 & 4 & 0 \\ 1 & 1 & 2 & 2 & 2 & 0 & 2 & 2 & 0 & 0 & 0 & 4 \end{array}\right)$$

The matrix can be partitioned into four 6×6 matrices with each of these matrices partitioned into four 3×3 matrices, as indicated. This shows the block circulant structure of the matrix. Thus

$$\mathbf{NN}' = \begin{pmatrix} \mathbf{B}_0 & \mathbf{B}_1 \\ \mathbf{B}_1 & \mathbf{B}_0 \end{pmatrix}$$

where $\mathbf{B}_h$ is a 6×6 matrix ($h = 0, 1$). Hence

$$\mathbf{NN}' = \begin{pmatrix} 1 & 0 \\ 0 & 1 \end{pmatrix} \otimes \mathbf{B}_0 + \begin{pmatrix} 0 & 1 \\ 1 & 0 \end{pmatrix} \otimes \mathbf{B}_1 = \sum_{h_1=0}^{1} (\mathbf{\Gamma}_{h_1} \otimes \mathbf{B}_{h_1})$$

where $\mathbf{\Gamma}_{h_i}$ is the $m_i \times m_i$ circulant matrix whose first row has 1 in the $(h_i + 1)$th column and zero elsewhere. Further, each $\mathbf{B}_{h_1}$ matrix can be written as

$$\mathbf{B}_{h_1} = \begin{pmatrix} \mathbf{B}_{h_1 0} & \mathbf{B}_{h_1 1} \\ \mathbf{B}_{h_1 1} & \mathbf{B}_{h_1 0} \end{pmatrix} = \sum_{h_2=0}^{1} (\mathbf{\Gamma}_{h_2} \otimes \mathbf{B}_{h_1 h_2})$$

where $\mathbf{B}_{h_1 h_2}$ is a 3×3 matrix. Finally, each $\mathbf{B}_{h_1 h_2}$ matrix is circulant, so that

$$\mathbf{B}_{h_1 h_2} = \sum_{h_3=0}^{2} \lambda_{h_1 h_2 h_3} \mathbf{\Gamma}_{h_3}$$

where the $\lambda_{h_1 h_2 h_3}$ are elements of the first row of $\mathbf{B}_{h_1 h_2}$. Therefore, the concurrence matrix for the above 3-cyclic design can be written as

$$\mathbf{NN}' = \sum_{h_1=0}^{1} \mathbf{\Gamma}_{h_1} \otimes \left(\sum_{h_2=0}^{1} \mathbf{\Gamma}_{h_2} \otimes \sum_{h_3=0}^{2} \lambda_{h_1h_2h_3} \mathbf{\Gamma}_{h_3} \right)$$

i.e.

$$\mathbf{NN}' = \sum_{h_1=0}^{1} \sum_{h_2=0}^{1} \sum_{h_3=0}^{2} \lambda_{h_1h_2h_3} (\mathbf{\Gamma}_{h_1} \otimes \mathbf{\Gamma}_{h_2} \otimes \mathbf{\Gamma}_{h_3})$$

where

$$r = \lambda_{000} = 4, \lambda_{001} = \lambda_{002} = 0, \lambda_{010} = 0, \lambda_{011} = \lambda_{012} = 2,$$
$$\lambda_{100} = 0, \lambda_{101} = \lambda_{102} = 2, \lambda_{110} = 2, \lambda_{111} = \lambda_{112} = 1$$

More generally, the block circulant concurrence matrix of an n-cyclic design can be written as

$$\mathbf{NN}' = \sum_{h_1=0}^{m_1-1} \sum_{h_2=0}^{m_2-1} \cdots \sum_{h_n=0}^{m_n-1} \lambda_{h_1h_2\ldots h_n} (\mathbf{\Gamma}_{h_1} \otimes \mathbf{\Gamma}_{h_2} \otimes \cdots \otimes \mathbf{\Gamma}_{h_n}) \quad (4.8)$$

By an extension of the methods used to obtain (4.2), explicit expressions can be obtained for the canonical efficiency factors of an n-cyclic design. They are

$$e_{u_1u_2\ldots u_n} = 1 - (1/rk) \sum_{h_1=0}^{m_1-1} \cdots \sum_{h_n=0}^{m_n-1} \lambda_{h_1\ldots h_n} \cos\left[\sum_{j=1}^{n} (2\pi u_j h_j / m_j) \right] \quad (4.9)$$

for $u_l = 0, 1, \ldots, m_l - 1$; $l = 1, 2, \ldots, n$. Note that if $n = 1$ then (4.9) is equivalent to (4.2).

4.6 Generalized cyclic designs

Generalized cyclic incomplete block designs were first considered by Jarrett and Hall (1978). The $v = mn$ treatments are divided into m (> 1) groups of n elements using the residue classes, modulo m. Thus, the ith residue class is

$$S_i = \{i, i+m, \ldots, i+m(n-1)\} \qquad (i = 0, 1, \ldots, m-1) \quad (4.10)$$

The designs are obtained by successive addition of m, modulo v, to the elements of one or more initial blocks. In general, each initial block contributes n blocks to the design, although again partial sets can be constructed. A generalized cyclic design with increment number m will be noted by GCIB_m.

Example 4.3

Suppose there are $v = 8$ treatments divided into 2 groups of 4. The residue classes are

$$S_0 = \{0, 2, 4, 6\}$$
$$S_1 = \{1, 3, 5, 7\}.$$

Two GCIB_2 designs for $k = 4$ and $b = 4$ can be obtained from the initial blocks (0 1 2 4) and (0 1 2 5). They are

$$\begin{array}{cccc} (0 & 1 & 2 & 4) \\ (2 & 3 & 4 & 6) \\ (4 & 5 & 6 & 0) \\ (6 & 7 & 0 & 2) \end{array} \qquad \begin{array}{cccc} (0 & 1 & 2 & 5) \\ (2 & 3 & 4 & 7) \\ (4 & 5 & 6 & 1) \\ (6 & 7 & 0 & 3) \end{array}$$

In the first design the treatments from class S_0 are each replicated three times while those from class S_1 are replicated once. In the second design all treatments are replicated twice.

It follows from the method of construction that all treatments within a residue class are equally replicated, although different residue classes may have different replications. Equal replication of all treatments is achieved if the initial blocks contain the same number of treatments from each residue class. For equal block size designs, the concurrence matrix $\mathbf{NN}'$ is completely determined from the first m rows. Hence, $\mathbf{NN}'$ can be partitioned into $m \times m$ submatrices, $\mathbf{B}_h(h = 1, 2, \ldots, n)$ arranged in a circulant manner, thus

$$\mathbf{NN}' = \sum_{h=0}^{n-1} \mathbf{\Gamma}_h \otimes \mathbf{B}_h$$

where $\mathbf{\Gamma}_h$ is a basic circulant matrix of order n as defined in the previous section. For instance, for the equal replicated design for $v = 8$ given in Example 4.3,

$$\mathbf{NN}' = \left(\begin{array}{cc|cc|cc|cc} 2 & 1 & 1 & 1 & 0 & 1 & 1 & 1 \\ 1 & 2 & 1 & 0 & 1 & 2 & 1 & 0 \\ \hline 1 & 1 & 2 & 1 & 1 & 1 & 0 & 1 \\ 1 & 0 & 1 & 2 & 1 & 0 & 1 & 2 \\ \hline 0 & 1 & 1 & 1 & 2 & 1 & 1 & 1 \\ 1 & 2 & 1 & 0 & 1 & 2 & 1 & 0 \\ \hline 1 & 1 & 0 & 1 & 1 & 1 & 2 & 1 \\ 1 & 0 & 1 & 2 & 1 & 0 & 1 & 2 \end{array}\right)$$

so that

$$\mathbf{B}_0 = \begin{pmatrix} 2 & 1 \\ 1 & 2 \end{pmatrix}, \quad \mathbf{B}_1 = \mathbf{B}_3 = \begin{pmatrix} 1 & 1 \\ 1 & 0 \end{pmatrix} \quad \text{and} \quad \mathbf{B}_2 = \begin{pmatrix} 0 & 1 \\ 1 & 2 \end{pmatrix}.$$

The matrix **A** of (1.12) and its generalized inverse matrix $\boldsymbol{\Omega}$ given by (1.17) or (1.37) will have the same structure. A similar structure also holds for the more general designs with unequal block sizes. It is not possible to write down explicit expressions for the efficiency factors of generalized cyclic designs in the way that was done for cyclic and n-cyclic designs. Jarrett and Hall (1978) show, however, that the canonical efficiency factors can be obtained by calculating the non-zero eigenvalues of certain $m \times m$ complex matrices.

As has already been indicated in Section 4.4, when $r < k$ cyclic designs will often be relatively inefficient in terms of the overall efficiency factor. In such cases it is frequently possible to find more efficient generalized cyclic designs. For example, consider a comparison of the GCIB_2 design given in Example 4.3 for $v = 8$ and initial block (0 1 2 5) with the partial cyclic set of the same size with initial block (0 1 4 5). In the GCIB_2 design the number of pairs of treatments occurring together in 0, 1 and 2 blocks are 6, 20 and 2 respectively. For the partial cyclic set, the frequencies of these concurrences are 8, 16 and 4 respectively. Using the counting rules Section 2.5, the GCIB_2 design will be preferable in that it has concurrences which are more nearly equal; the average efficiency factor is 0.808 for the GCIB design compared with 0.778 for the cyclic design. The incidence matrix for this GCIB_2 design is

$$\mathbf{N} = \begin{pmatrix} 1 & 0 & 0 & 1 \\ 1 & 0 & 1 & 0 \\ 1 & 1 & 0 & 0 \\ 0 & 1 & 0 & 1 \\ 0 & 1 & 1 & 0 \\ 1 & 0 & 1 & 0 \\ 0 & 0 & 1 & 1 \\ 0 & 1 & 0 & 1 \end{pmatrix}$$

Note that the odd-numbered rows of **N** are obtained by cyclic development of the first row and that the even-numbered rows from a cyclic development of the second row. Thus the matrix $\mathbf{N}'$ is the incidence matrix of two full cyclic sets with initial blocks given by the first two rows of **N**. The dual of this $GCIB_2$ design is, therefore, the cyclic design for four treatments with initial blocks (0 3) and (0 2), with the half-set (0 2) replicated twice.

In general, the dual of a $GCIB_m$ design for $v = mn$ treatments in n blocks of k is a cyclic design with n treatments in v blocks of $r = k/m$ obtained from m initial blocks. It follows from (2.27), therefore, that efficient cyclic designs can be used to construct efficient generalized cyclic designs. Using the best cyclic design does not, however, necessarily produce the best generalized cyclic design for, as Hall and Jarrett (1981) have shown, better designs may be obtained in some cases by increasing the value of m. For example, consider the construction of a generalized cyclic design for $v = 32, r = 2, k = 4$ and $b = 16$. With $m = 2$ and $n = 16$, the best cyclic design for 16 treatments in 32 blocks of 2 has initial blocks (0 1) and (0 6). The dual of this design is the $GCIB_2$ design with initial block (0 1 2 13) and average efficiency factor $E = 0.619$. A more efficient generalized cyclic design of the same size is obtained from the initial blocks (0 1 2 5) and (0 3 11 14) with increment number $m = 4$. The average efficiency factor of this design is $E = 0.624$.

The above result on the duality of cyclic designs is a special case of a more general result given by Hall and Jarrett (1981). They show that the dual of a $GCIB_m$ design with mn treatments and qn blocks is a $GCIB_q$ design with qn treatments and mn blocks. They use this result to produce tables of efficient generalized cyclic designs for $10 \leqslant v \leqslant 60$, $r \leqslant k$ and for $(r, k) = (2, 4), (2, 6), (2, 8), (2, 10), (3, 6), (3, 9), (4, 6), (4, 8), (4, 10), (5, 10)$. For other cases with $r \leqslant 5$ and $k \leqslant 10$, r and k have no factors in common and k must necessarily divide v. Generalized cyclic designs with equal replication are then equivalent to the α-designs of Patterson and Williams (1976a), which have been extensively catalogued and which are discussed in Section 4.8.

4.7 Resolvable cyclic and generalized cyclic designs

In a resolvable design the $v = ks$ treatments are set out in r groups of s blocks each having k plots. Within each group each treatment is replicated once. The lattice designs considered in Section 3.4 are only available for limited numbers of treatments and block sizes. Square lattices have $k = s$, while for rectangular lattices $k = s - 1$. A large number of resolvable designs can be obtained by cyclical methods of construction. These methods will be discussed in the

Table 4.5 *Resolvable cyclic design*

Replication group												
	1				*2*			...		*k*		
0	k	...	$v-k$	1	$k+1$	...	$v-k+1$	...	$k-1$	$2k-1$	...	$v-1$
1	$k+1$	...	$v-k+1$	2	$k+2$	...	$v-k+2$	...	k	$2k$	...	0
⋮	⋮		⋮	⋮	⋮		⋮		⋮	⋮		⋮
$k-1$	$2k-1$	...	$v-1$	k	$2k$	...	0	...	$2k-2$	$3k-2$	...	$k-2$

remainder of this chapter. It should be noted that some other incomplete block designs are also resolvable. Information on the resolvability of group-divisible designs and other partially balanced incomplete block designs with two associate classes is given in the catalogue by Clatworthy (1973). Such designs are, however, usually lattice designs or less efficient alternatives.

David (1967) shows that the cyclic designs of Section 4.2 can be used to construct resolvable designs. One such series of designs is given by the set with initial block (0 1 2...$k-1$) since it may be written in k groups as in Table 4.5, where for convenience columns represent blocks. Such sets, however, tend to be inefficient since the elements in the concurrence matrix $\mathbf{NN}'$ are usually very unequal; see for instance Example 4.1. Since, in any group, the treatments in a block can be obtained by adding k to the treatments in the previous block, it follows that a cyclic set is resolvable if the initial block is of the form (0 1 2...$k-1$) after each label has been reduced modulo k. Many of the sets will have $r=k$ but partial sets with $r<k$ also exist. For instance, for $v=12$ and $k=4$ the sets with initial blocks (0 1 2 3), (0 1 3 6), (0 1 6 7) and (0 3 6 9) are resolvable. The first two sets have $r=k=4$, the third $r=2$ and the last $r=1$. Designs with $r>k$ can be constructed by suitable combinations of sets.

The above requirement for the form of the initial block is both necessary and sufficient when $r=k$, but is not necessary when $r<k$. For example, for $v=8$ and $k=4$ the initial block (0 1 4 5) gives a resolvable partial set with $r=2$ but is not of the form (0 1 2 3) after reduction modulo 4. Patterson and Williams (1976a) define a more general series of resolvable cyclic designs for integral k/r. Now the initial block consists of the k possible sums of each element of the

set $(0, b, 2b, \ldots, (k-r)b/r)$ with each element of a set which is of the form $(0\ 1\ 2 \ldots r-1)$ after reduction modulo r, where $b = rs$ is the total number of blocks. The design for $v = 8$ above is included in this series since the initial block (0 1 4 5) is obtained from the sets (0 4) and (0 1).

It is clear that the resolvable cyclic design with initial block $(0\ 1\ 2 \ldots k-1)$ is also a GCIB_k design with k initial blocks given by $(i-1\ i\ i+1 \ldots k+i-2)$, $i = 1, 2, \ldots, k$. Every initial block contains one treatment from each of the k residue classes. More generally, the successive addition of k modulo v to an initial block containing one treatment from each residue class implies that each initial block generates one replicate. Generalized cyclic designs, therefore, provide a very flexible class of resolvable designs. With a suitable relabelling of treatments, however, these designs are equivalent to the α-designs of Patterson and Williams (1976a), which are considered in detail in the next section.

4.8 α-Designs

4.8.1 *Construction*

Although cyclic designs provide a large number of resolvable designs in addition to the lattice designs, there are still a number of combinations of v and r for which no such design exists or where the existing design is of low efficiency. A more general series of resolvable designs for $v = ks$ treatments, called α-designs, has been given by Patterson and Williams (1976a). For these designs there is no limitation on block size other than the unavoidable constraint that v/k must be an integer.

The construction of an α-design starts with a $k \times r$ array α whose elements are in the set of residues mod s. Each column of α is used to generate $s-1$ further columns by cyclic substitution. The resulting $k \times rs$ array is denoted by α^*. Finally, s is added to each element in the second row of α^*, $2s$ is added to each element in the third row, and so on. The columns of the resulting array are now the blocks of the required design and each set of columns generated from the same column of α constitutes a complete replication.

Each column of the generating array α can also be regarded as an initial block of a GCIB_k design if element j in the ith row of the array is replaced by $kj + (i-1)$.

Example 4.4

Consider the construction of an α-design for three replications of 12 treatments, each replication consisting of three blocks of four plots, i.e. $v = 12, k = 4, r = s = 3, b = 9$. Given the generating array α below, the intermediate array α^* and, hence, the design are easily constructed.

Generating array α

0	0	0
0	0	2
0	2	1
0	1	1

*Intermediate array α**

0	1	2	0	1	2	0	1	2
0	1	2	0	1	2	2	0	1
0	1	2	2	0	1	1	2	0
0	1	2	1	2	0	1	2	0

Design

Replicate	1			2			3		
Block	1	2	3	4	5	6	7	8	9
	0	1	2	0	1	2	0	1	2
	3	4	5	3	4	5	5	3	4
	6	7	8	8	6	7	7	8	6
	9	10	11	10	11	9	10	11	9

This design is equivalent to the $GCIB_4$ design given by the initial blocks (0 1 2 3), (0 1 10 7) and (0 9 6 7).

From the results in Section 4.6, it follows that the dual of an α-design is an α-design. Williams (1975a) shows that the generating array of the dual design is given by transposing the rows and columns of the generating array of the original design; the transposed array being denoted by α'. This also follows from results given by Hall and Jarrett (1981). It can be verified that the dual of the design in Example 4.4 is the resolvable $GCIB_3$ design for $v = 9, k = s = 3, r = 4$ with initial blocks (0 1 2), (0 1 8), (0 7 5) and (0 4 5).

4.8.2 Concurrences

Different α-designs can be obtained from different generating arrays, and the choice of an appropriate design can again be based on the efficiency factors of the designs. The problem of choosing optimal designs can be overcome by a combination of theory and computing. The theory can be used to identify classes of designs with high efficiency factors. These designs can then be generated on a computer and the best catalogued.

The efficiency factors of a design of a given size depend on the off-diagonal elements of the concurrence matrix $\mathbf{NN}'$. The search for optimal α-designs is simplified by the fact that the number of concurrences of any two treatments, i.e. the number of blocks containing both treatments, can be determined directly from the generating array α. It is not necessary to construct the full design. Let $\alpha(p, q)$ be the (pq)th element of the array α $(p = 1, 2, \ldots, k; q = 1, 2, \ldots, r)$. Then, the number of concurrences of treatments i and j is the frequency of $(j - i) \bmod s$ in the set of r differences $\{\alpha(p_j, q) - \alpha(p_i, q)\} \bmod s$ $(q = 1, 2, \ldots, r)$, where p_i is one more than the integer part of i/s.

As an illustration, consider the concurrence of treatments 1 and 8 in the design for 12 treatments in blocks of 4 given in Example 4.4. The values of p_i, p_j and $(j - i) \bmod 3$ are 1, 3 and 1 respectively. The elements of the first j rows of α have, therefore, to be subtracted from those in the third row. The three differences are 0, 2, 1 and since 1 occurs once, then treatments 1 and 8 concur once, namely in the third replicate. It follows immediately that the pairs 0 and 7 and 2 and 6 also concur once. Other pairs of treatments also concur once but some concur twice and others not at all.

A design with concurrences $g_1, g_2, \ldots$ will be referred to as an $\alpha(g_1, g_2, \ldots)$-design; the example above is therefore an $\alpha(0, 1, 2)$-design.

4.8.3 *Reduced array*

The generating array in Example 4.4 is of a special kind, called a *reduced* array, with all elements equal to zero in the first row and column. Under a suitable relabelling or permutation of elements all arrays can be represented as reduced arrays, thereby restricting the search for optimal designs to reduced arrays only. For example, consider the following generating array for $s = 3$:

$$\begin{matrix} 1 & 0 & 1 \\ 2 & 1 & 1 \\ 0 & 1 & 1 \\ 1 & 1 & 2 \end{matrix}$$

First add 2 to the elements in column 1, and 2 to those in column 3, in each case reducing modulo 3 when necessary. This results in a rearrangement of blocks and gives the array

$$\begin{matrix} 0 & 0 & 0 \\ 1 & 1 & 0 \\ 2 & 1 & 0 \\ 0 & 1 & 1 \end{matrix}$$

Now add 2 to the elements in row 2, and 1 to those in row 3, and again reduce modulo 3 when necessary. This operation simply relabels the treatments and the resulting array is the reduced array used in Example 4.4 to give the $\alpha(0,1,2)$-design for 12 treatments.

4.8.4 Choice of design

Tables of suitable generating arrays have been given by Williams (1975a) for $v \leqslant 100$ and $r = 2, 3, 4$. They have been constructed with the aim of choosing, for every admissible value of k, a single α-design with efficiency factor E as large as possible, where E is given by (2.9). Designs with maximum E among all α-designs are called α-optimal. For $r = 2$, as will be shown in Section 4.9, the α-optimal designs can be obtained from the most efficient symmetric cyclic designs for s treatments in blocks of k. For a few combinations of v, k and $r = 3$ or 4, the α-optimal designs are known, for instance when they correspond to square lattices. In general, however, the tabulated designs were obtained by first selecting a set of non-isomorphic designs likely to be of high efficiency and then calculating and comparing the E values of these designs. The upper bound for E for resolvable designs given in (3.15) is useful in judging whether there is room for further improvement over the α-designs listed.

Following the criteria for choosing efficient designs given in Section 2.5, the set of α-designs with high efficiency are those which minimize the range of off-diagonal elements in the concurrence matrix. Thus, given the choice, an $\alpha(0,1)$-design would be preferred to an $\alpha(0,1,2)$-design. If an $\alpha(0,1,2)$-design has to be used, then one which has as few pairs of treatments as possible concurring twice would be preferred.

From the construction of α-designs it is clear that neither balanced α-designs nor $\alpha(1,2)$-designs can exist, since the concurrence of any two treatments i and j such that the integer parts of i/s and j/s are

Table 4.6 *Basic generating arrays for α-designs.*

$s = k = 5$

0	0	0	0	0
0	1	2	3	4
0	4	3	2	1
0	2	4	1	3

$s = k = 10$

0	0	0	0	0	0	0	0	0	0
0	1	3	5	4	6	7	8	9	2
0	9	6	7	5	3	2	4	8	6
0	5	9	2	6	1	4	7	2	3

$s = k = 6$

0	0	0	0	0	0
0	1	3	2	4	5
0	5	2	3	1	1
0	4	5	1	2	3

$s = 11, k = 9$

0	0	0	0	0	0	0	0	0
0	1	4	9	2	5	6	3	7
0	6	8	7	3	1	5	9	4
0	7	1	5	6	3	10	4	1

$s = k = 7$

0	0	0	0	0	0	0
0	1	2	4	3	5	6
0	3	6	5	2	1	4
0	2	4	1	6	3	5

$s = 12, k = 8$

0	0	0	0	0	0	0	0
0	1	7	9	4	11	10	5
0	2	5	6	11	3	4	1
0	3	1	4	8	10	7	6

$s = k = 8$

0	0	0	0	0	0	0	0
0	1	3	5	2	4	6	7
0	2	7	3	5	1	0	6
0	6	1	4	3	6	2	5

$s = 13, k = 7$

0	0	0	0	0	0	0
0	1	3	9	12	8	6
0	4	8	2	10	5	7
0	10	11	1	6	12	8

$s = k = 9$

0	0	0	0	0	0	0	0	0
0	1	3	7	2	4	5	6	8
0	8	6	2	3	1	7	5	4
0	7	4	3	5	6	2	1	7

$s = 14, k = 7$

0	0	0	0	0	0	0
o	1	9	11	2	5	3
0	8	10	13	6	11	1
0	10	7	2	1	12	11

$s = 15, k = 6$

0	0	0	0	0	0
0	1	3	7	10	14
0	8	12	2	13	3
0	7	14	5	11	8

Reproduced from Patterson, Williams and Hunter (1978), Block designs for variety trials. *J. Agric. Sci.*, **90**, p. 399, by permission of Cambridge University Press.

equal is impossible. For instance, in the design for 12 treatments in Example 4.4 it is not possible for treatments 0 and 1 to occur together in any block.

A necessary, though not sufficient, condition for the existence of an $\alpha(0,1)$-design is that $k \leqslant s$. Williams (1975a) lists an $\alpha(0,1)$-design for 188 of the 198 combinations of v, r and k satisfying $k \leqslant s$, $r = 2$, 3, 4 and $v \leqslant 100$. Of the remaining 10 cases, lattice designs exist for nine of them while an $\alpha(0,1,2)$-design has to be used for the case with $s = k = 6$ and $r = 4$.

The tables of designs given by Williams (1975a) are, however, large and fairly inaccessible. As an alternative, Patterson, Williams and Hunter (1978) have given 11 basic arrays which can be used to provide α-designs with $k \leqslant s$. These arrays are reproduced in Table 4.6, where the generating arrays have for convenience been written as $r \times k$ arrays, rather than $k \times r$ arrays.

The arrays are for four-replicate designs with k equal to the smaller of s and the integer part of $100/s$, and s ranging from 5 to 15. Arrays for other values of $k \geqslant 4$ are obtained from the first k columns of the basic arrays and the arrays for $r = 2$, 3 from the first two or three rows. Table 4.6 must not be used to obtain designs with $k < 4$.

Patterson et al. (1978) state that 77 of the 147 α-designs given in Table 4.6 have efficiency factors at least as large as any other resolvable design with the same v, s, k and r. The remainder have efficiency factors falling short of efficiency factors of known alternatives by varying amounts. However, if square and rectangular lattices are used whenever possible and if the generating array

$$\begin{array}{ccccc} 0 & 0 & 0 & 0 & 0 \\ 0 & 1 & 2 & 4 & 7 \end{array}$$

is used for the two cases $r = 2, k = 5, s = 11$ and 12, the largest shortfall in efficiency factors is only .0022 and unlikely to be of practical importance. Table 4.6, therefore, provides an effective and compact alternative to the very best designs listed in the extensive tables given by Williams (1975a). For $k > s$, the best $\alpha(0,1,2)$-designs for $r = 2, 3, 4$ and $v \leqslant 100$ are listed in Williams (1975a).

A further alternative to these tables of designs is the computer algorithm for choosing suitable generating arrays described by Paterson and Patterson (1983). The choice of array is based on the aim of minimizing the number of circuits of lengths 3 and 4 in an α-design; the use of circuits to obtain efficient block designs has been discussed in Section 2.5. One obvious benefit of the algorithm is that it provides a method of obtaining efficient α-designs for parameter values not included in the tables of Williams.

4.8.5 *Unequal block sizes*

Resolvable designs with equal sized blocks of k plots with $k < v$ can be used only when v is a multiple of k. In many practical cases a factorization of the number of treatments in the form $v = ks$ may not be possible. If resolvability is still required then it becomes necessary to use unequal block sizes. Patterson and Williams (1976a) consider the use of two block sizes k_1 and k_2 with the inequality in block size minimized by imposing the condition that $k_2 = k_1 - 1$. Hence, these designs exist when v can be expressed in the form $v = s_1k_1 + s_2k_2$, where s_1, s_2, k_1 and k_2 are positive integers. The designs are derived from α-designs as follows:

(i) Construct an α-design for $v + s_2$ treatments with $s = s_1 + s_2$ blocks of k_1 plots in each replication.
(ii) Delete a set of s_2 treatments, no two of which concur. The varieties labelled $v, \ldots, v + s_2 - 1$ provide such a set.

For example, deletion of the treatments labelled 10 and 11 from the design in Example 4.4 gives a design for 10 treatments with one block of 4 plots and two blocks of 3 plots in each of three replications.

4.9 Two-replicate resolvable designs

Designs with two replications are important in practice when economy is required in the amount of experimental material (seed, for example) or in the number of plots. Patterson and Williams (1976b) present results which enable the optimal resolvable designs for $r = 2$ to be determined. They show that every binary resolvable incomplete block design for $v = ks, b = 2s$ and $r = 2$ is uniquely determined by a symmetric block design with $v = b = s$ and $r = k$. This symmetric design is called the *contraction* of the resolvable design.

An example will illustrate the general approach to establishing this result. Consider the following two-replicate resolvable design for 12 treatments in 8 blocks of 3, where the two replicates are labelled 0 and 1 and, within each replicate, blocks are labelled 0, 1, 2, 3.

Replicate	0				1			
Block	0	1	2	3	0	1	2	3
	0	1	2	3	0	1	2	3
	4	5	6	7	5	6	7	4
	8	9	10	11	10	11	8	9

The *block concurrence graph* for this design is shown in Fig. 4.1. A block concurrence graph, based on the block concurrence matrix $\mathbf{N}'\mathbf{N}$, has the blocks of the design as points, and the lines correspond to treatments which pairs of blocks have in common. For instance, there is a line between block 1 in replicate 0 and block 3 in replicate 1, showing that these two blocks have a treatment in common, namely treatment 9.

The graph can also be regarded as a *design* graph. Now the four upper points correspond to four treatment labels and the four lower points to four block labels. If a line joins a block point to a treatment point then the block will contain that treatment. Hence, the design graph produces the symmetric block design (contraction)

(0 1 2), (1 2 3), (2 3 0), (3 0 1).

The same graph can, therefore, be regarded as giving rise to both the resolvable design and its contraction, hence establishing the corresponding between them. Patterson and Williams (1976b) proved

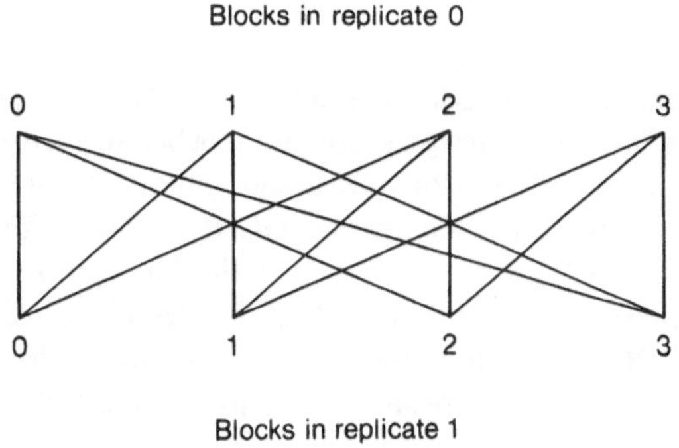

Figure 4.1 *The block concurrence graph.*

that the average efficiency factor E_R of the resolvable design is given by

$$E_R = \frac{ks - 1}{ks - 2s + 1 + 4(s-1)E_s^{-1}} \tag{4.11}$$

where E_s is the average efficiency factor of the contraction.

An immediate consequence of (4.11) is that the resolvable design is A-optimal if the contraction is A-optimal. The above contraction with four treatments, for example, is a balanced incomplete block design (it is also a cyclic design) so that the resolvable design given for 12 treatments in 8 blocks of 3 is A-optimal.

If the contraction is a complete block design ($k = s$) the resulting two-replicate resolvable design is a simple lattice, which is therefore A-optimal. If the contraction is a balanced incomplete block design the resolvable design is also A-optimal; included here are the simple rectangular lattices. If $k > s$ the contraction will be non-binary. Consideration of the counting rules given in Section 2.5 shows that the incidence matrix $\mathbf{N}$ of an efficient non-binary design will be of the form $\mathbf{N} = \mathbf{N}_0 + p\mathbf{J}$, for some integer $p > 1$ and where $\mathbf{N}_0$ is the incidence matrix of a binary design, i.e. within each block some treatments are replicated p times while the others are replicated $p + 1$ times. Again, if $\mathbf{N}_0$ is the incidence matrix of a balanced incomplete block design the resolvable design will be A-optimal.

If the contraction is a symmetric cyclic design then the two-replicate resolvable design will be an α-design. Hence, as stated in Section 4.8.4, the α-optimal designs are obtained from the most efficient symmetric cyclic designs for s treatments in blocks of k. The cyclic designs required for this purpose can be obtained from the tables referred to in Section 4.4.

Suppose that the cyclic contraction has initial block $(0\, a_1\, a_2 \cdots a_{k-1})$ and that the lines in its design graph are labelled in such a way that the first replicate of the resolvable design, obtained from the block concurrence graph, is given by the GCIB_k set with initial block $(0\ 1\ 2\ \cdots\ k-1)$. Since the treatment in the jth position of the a_ith block of this set is $ka_i + j$, it follows that the second replicate is given by the GCIB_k set with initial block $(0\ b_1\ b_2\ \cdots\ b_{k-1})$, where $b_i = ka_i + i$. Hence, from results given in Section 4.8.1, the two-replicate resolvable design is an α-design obtained from the generating array

$$\begin{matrix} 0 & 0 \\ 0 & a_1 \\ 0 & a_2 \\ \vdots & \vdots \\ 0 & a_{k-1} \end{matrix}$$

Example 4.5

Consider the construction of a two-replicate α-design for 54 treatments in blocks of 9. The most efficient symmetric cyclic set for 6 treatments in blocks of 3 has initial block (0 1 3), so that an efficient non-binary cyclic contraction for $s = 6$ and $k = 9$ is obtained from the initial block (0 1 2 3 4 5 0 1 3). Hence, the generating array of the α-design is given by the transpose of

$$\begin{matrix} 0 & 0 & 0 & 0 & 0 & 0 & 0 & 0 & 0 \\ 0 & 0 & 1 & 1 & 2 & 3 & 3 & 4 & 5 \end{matrix}$$

The full α-design can now be set out following the method of construction given in Section 4.8.1. Alternatively, it is the GCIB_9 design with initial blocks

(0 1 2 3 4 5 6 7 8), (0 1 11 12 22 32 33 43 53)

Further results on the construction of optimal two-replicate resolvable designs can be found in Williams, Patterson and John (1976, 1977).

4.10 Resolvable paired-comparison designs

The dual of a two-replicate α-design is a resolvable paired-comparison ($k = 2$) design. In fact, Williams (1976) first established the link between the average efficiency factor E_R of an α-design with $k = 2, v = 2s$ and generating array

$$\begin{matrix} 0 & 0 & \cdots & 0 \\ 0 & a_1 & \cdots & a_{r-1} \end{matrix}$$

and the average efficiency factor E_s of a cyclic design for s treatments with initial block $(0\ a_1\ \cdots\ a_{r-1})$. Using (4.11) and (2.27) it can be shown that E_R is given by (4.11) with $k = 2$. Hence, α-optimal designs for $k = 2$ are obtained from the most efficient cyclic designs. Many A-optimal resolvable paired-comparison designs can be constructed this way as many of the A-optimal block designs are cyclic.

For $6 \leqslant v \leqslant 14$, $r \leqslant s$ and $k = 2$, Williams (1976) compared the best α-designs with the best resolvable cyclic designs given by David (1967). In 18 of the 20 cases considered the α-designs were either better than or equivalent to the cyclic designs. Williams also showed that the best α-designs compare favourably with the paired comparison designs listed in Table A of the catalogue of cyclic designs by John, Wolock and David (1972), even though these cyclic designs are not necessarily resolvable.

CHAPTER 5

Row–column designs

5.1 Introduction

In row–column designs the experimental units are grouped in two directions, i.e. two blocking factors are used with one factor representing the rows of the design and the other factor representing columns. For instance, in an agricultural experiment the rows and columns of a design might represent different times and method of application of fertilizers. In a preference testing experiment, the columns might represent different subjects and the rows the order in which different products are presented to the subjects. In an insulin-response experiment, the rows, columns and treatments may represent respectively individual animals, date of injection and levels of insulin. In all these experiments it is hoped that there will be a gain in the accuracy of estimating treatment comparisons resulting from eliminating the effects of the row and column factors.

More generally, an experiment may involve blocks of row–column designs. These blocks may represent a further blocking factor or provide replications of the basic experimental design. For example, the above row–column agricultural experiment may be repeated in various locations.

The designs considered in this chapter have v treatments set out in b blocks of $k = pq$ units per block. The treatments within each block are then arranged in a row–column design consisting of p rows and q columns. For example, a (Latin square) design for $v = 4$ treatments in a single block ($b = 1$) of $p = 4$ rows and $q = 4$ columns is given by

0	1	2	3
1	3	0	2
2	0	3	1
3	2	1	0

Table 5.1 *Partition of degrees of freedom in a row–column design.*

Stratum	*d.f.*
Between blocks	$b-1$
Between rows (within blocks)	$b(p-1)$
Between columns (within blocks)	$b(q-1)$
Within rows and columns (within blocks)	$b(p-1)(q-1)$
Total	$bpq-1$

A (lattice square) design for $v=9$ treatments in $b=2$ blocks each containing $p=3$ rows and $q=3$ columns is given by

0	1	2		0	4	8
3	4	5		5	6	1
6	7	8		7	2	3

A design based on incomplete blocks ($k < v$) with $v=8$, $b=4$, $p=2$ and $q=3$ is given by

2	6	3		1	5	2		3	7	4		0	4	1
4	7	0		3	6	7		5	0	1		2	5	6

The degrees of freedom in a row–column design can be partitioned into four blocking strata as shown in Table 5.1. Information on treatment comparisons may be available in all strata, though comparisons made in the lowest stratum, namely the within rows and columns stratum, will usually have the greatest precision. The aim in choosing an appropriate row-column design will, therefore, be to maximize the amount of information available on treatment comparisons in this stratum.

5.2 Model and information matrix

Let y_{ijlm} be the response obtained from allocating the mth treatment to the jth row and lth column of the ith block. Then the model corresponding to the partition given in Table 5.1 is

$$y_{ijlm} = \mu + \beta_i + \rho_{ij} + \gamma_{il} + \tau_m + \varepsilon_{ijlm} \tag{5.1}$$

$$(i=1, 2, \ldots, b;\ j=1, 2, \ldots, p;\ l=1, 2, \ldots, q;\ m=1, 2, \ldots, v)$$

where the β_i, ρ_{ij} and γ_{il} parameters represent respectively block, row (within-block) and column (within-block) effects, and τ_m represents the effect of applying the mth treatment. The error terms ε_{ijlm} are assumed to be uncorrelated random variables each with mean zero and variance σ^2.

Let D_b denote the design used in setting out the v treatments in b blocks of $k = pq$ units per block. Let D_p and D_q similarly denote the block designs given by the rows and columns respectively of the row–column design. These three designs are the *component* designs of the row–column design. For instance, in the lattice square design for $v = 9$ treatments given in the previous section, D_b is a randomized block design with 2 blocks and D_p and D_q can both be shown to be (Latin square type) PBIB/2 designs (see Section 3.1) for 9 treatments in 6 blocks of 3 with $\lambda_1 = 1$ and $\lambda_2 = 0$.

Let $\mathbf{N}_b$, $\mathbf{N}_p$ and $\mathbf{N}_q$ be the incidence matrices and $\mathbf{A}_b$, $\mathbf{A}_p$ and $\mathbf{A}_q$ the information matrices of the designs D_b, D_p and D_q respectively. Similarly, let $\mathbf{q}_b, \mathbf{q}_p$ and $\mathbf{q}_q$ be the vectors of adjusted treatment totals for the three component designs. Thus, from (1.12) and (1.13),

$$\mathbf{A}_b = \mathbf{r}^{\delta} - (1/k)\mathbf{N}_b\mathbf{N}_b', \qquad \mathbf{q}_b = \mathbf{T} - (1/k)\mathbf{N}_b\mathbf{B}$$

with $k = pq$ replaced by q and p, and $\mathbf{B}$ replaced by $\mathbf{R}$ and $\mathbf{C}$ for designs D_p and D_q respectively, where $\mathbf{T}, \mathbf{B}, \mathbf{R}$ and $\mathbf{C}$ are respectively vectors of treatment, block, row and column totals. As before, $\mathbf{r}$ is the replication vector.

The reduced normal equations for the treatment parameters, after eliminating block, row and column parameters, can be obtained in a way analogous to the method used in Section 1.3 for block designs. For row–column designs these equations are

$$\mathbf{A}\hat{\boldsymbol{\tau}} = \mathbf{q} \tag{5.2}$$

where

$$\mathbf{A} = \mathbf{A}_p + \mathbf{A}_q - \mathbf{A}_b \tag{5.3}$$

and

$$\mathbf{q} = \mathbf{q}_p + \mathbf{q}_q - \mathbf{q}_b \tag{5.4}$$

The results given in Sections 1.3–1.5 for block designs can be applied in the same way to these row–column designs. For instance, a solution to (5.2) is given by $\hat{\boldsymbol{\tau}} = \boldsymbol{\Omega}\mathbf{q}$ where $\boldsymbol{\Omega}$ is a generalized inverse of $\mathbf{A}$; the estimable function $\mathbf{c}'\boldsymbol{\tau}$ satisfies $\mathbf{c}' = \mathbf{c}'\boldsymbol{\Omega}\mathbf{A}$ and has unbiased estimator $\mathbf{c}'\hat{\boldsymbol{\tau}}$ with variance $\mathbf{c}'\boldsymbol{\Omega}\mathbf{c}\sigma^2$; and the (adjusted) treatment sum of squares is $\hat{\boldsymbol{\tau}}'\mathbf{q}$.

5.3 Canonical efficiency factors

Following Section 2.7, let e_j be an eigenvalue corresponding to an eigenvector $\mathbf{s}_j$ of the matrix $\mathbf{A}$ given in (5.2) with respect to $\mathbf{r}^\delta$, i.e. $\mathbf{A}\mathbf{s}_j = e_j\mathbf{r}^\delta\mathbf{s}_j (j = 1, 2, \ldots, v)$. The non-zero eigenvalues $e_1, e_2, \ldots, e_m$ are the canonical efficiency factors of the row–column design, where $m = \text{rank}(\mathbf{A})$. Optimality criteria based on these factors can be used, as was done with block designs, to assess different row–column designs.

In general, the canonical efficiency factors are most easily obtained as the (non-zero) eigenvalues of the matrix $\mathbf{A}^* = \mathbf{r}^{-\delta/2}\mathbf{A}\mathbf{r}^{-\delta/2}$; see (2.24). However, one case of particular interest arises when the information matrices $\mathbf{A}_b, \mathbf{A}_p$ and $\mathbf{A}_q$ of the component designs are spanned by the same set of eigenvectors. Suppose

$$\mathbf{A}_b\mathbf{s}_j = e_{bj}\mathbf{r}^\delta\mathbf{s}_j, \qquad \mathbf{A}_p\mathbf{s}_j = e_{pj}\mathbf{r}^\delta\mathbf{s}_j, \qquad \mathbf{A}_q\mathbf{s}_j = e_{qj}\mathbf{r}^\delta\mathbf{s}_j \tag{5.5}$$

Then, from (5.3), the jth canonical efficiency factor e_j of the row–column design is given by

$$e_j = e_{pj} + e_{qj} - e_{bj} \tag{5.6}$$

For the equal replication case, (5.6) will hold if the component designs are complete block, balanced incomplete block (or any variance-balanced design), cyclic or n-cyclic designs. Cyclic and n-cyclic designs satisfy this requirement since circulant and block circulant matrices of a given size have a common set of eigenvectors. Note also that if any two of the component designs are variance-balanced (or efficiency-balanced in the unequal replicate case) then (5.6) will hold no matter what type of design is used for the remaining component.

As an example, for the lattice design for 9 treatments given in Section 5.1 the three information matrices $\mathbf{A}_b, \mathbf{A}_p$ and $\mathbf{A}_q$ have a common set of eigenvectors, since design D_b is a variance-balanced design with $e_{bj} = 1$ for all j and the matrices $\mathbf{A}_p$ and $\mathbf{A}_q$ are block circulants. The canonical efficiency factors for D_p and D_q, corresponding to these common set of eigenvectors, are

$$e_{pj} = .5, .5, .5, .5, 1, 1, 1, 1$$
$$e_{qj} = 1, 1, 1, 1, .5, .5, .5, .5$$

Hence, from (5.6), the lattice square design is variance-balanced with all canonical efficiency factors equal to 0.5. This result can be verified by showing that $\mathbf{A} = \mathbf{I}_9 - \mathbf{K}_9$ since it then follows from (2.11) that the design is variance-balanced.

5.4 Orthogonality and connectedness

In an orthogonal row–column design estimates of the treatment parameters will be the same as those obtained from the model with no block, row or column parameters. The requirements for orthogonality are, following the methods used in Section 1.7, that

$$\mathbf{N}_b = (1/b)\mathbf{r}\mathbf{1}'_b, \qquad \mathbf{N}_p = (1/pb)\mathbf{r}\mathbf{1}'_{pb}, \qquad \mathbf{N}_q = (1/qb)\mathbf{r}\mathbf{1}'_{qb} \tag{5.7}$$

The Latin square design for 4 treatments given in Section 5.1 is an orthogonal row–column design since $\mathbf{N}_b = 4\,\mathbf{1}\,\mathbf{1}'$, $\mathbf{N}_p = \mathbf{N}_q = \mathbf{1}\,\mathbf{1}'$ satisfy (5.7). The component designs D_p and D_q are both (orthogonal) randomized block designs as each treatment occurs once in each row and once in each column. For a more complex example, an orthogonal row–column design for $v = 2$ treatments with $b = 2$, $p = q = 3$ and $\mathbf{r} = (6, 12)'$ is

0	1	1	1	0	1
1	0	1	0	1	1
1	1	0	1	1	0

Using (5.7), the information matrix and vector of adjusted treatment totals for an orthogonal design are, from (5.3) and (5.4),

$$\mathbf{A} = \mathbf{r}^{\delta} - (1/n)\mathbf{r}\mathbf{r}' \tag{5.8}$$

and

$$\mathbf{q} = \mathbf{T} - \bar{y}\mathbf{r} \tag{5.9}$$

where $n = bpq$. These formulae are the same as for an orthogonal block design given in (1.40) and (1.41) so that the results of Section 1.7 apply here also. In particular, since $\boldsymbol{\Omega} = \mathbf{r}^{-\delta}$ is a generalized inverse of $\mathbf{A}$, estimates of the treatment parameters are given by the unadjusted treatment means.

Adjusted orthogonality in a multi-factor design has been considered by Eccleston and Russell (1975, 1977). Of particular importance in row–column designs is the concept of orthogonality between rows and columns after adjusting for treatments. This is defined as follows. Consider the model obtained from (5.1) by deleting the column parameters γ_{il}, namely

$$y_{ijlm} = \mu + \beta_i + \rho_{ij} + \tau_m + \varepsilon_{ijlm} \tag{5.10}$$

Now if the estimates of the row parameters ρ_{ij} are the same for both models (5.1) and (5.10) then rows are said to be adjusted orthogonal to columns. Similarly, adjusted orthogonality means that estimates of column parameters will be the same regardless of whether the row parameters are included in the model. For equal replicate designs, after deleting the treatment parameters from the normal equations for model (5.1) it can be shown that the necessary and sufficient condition for adjusted orthogonality between rows and columns is that

$$\mathbf{N}_p'\mathbf{N}_q = (r/b)\mathbf{J} \tag{5.11}$$

Now since $\mathbf{N}_b = \mathbf{N}_p(\mathbf{I}_b \otimes \mathbf{1}_p) = \mathbf{N}_q(\mathbf{I}_b \otimes \mathbf{1}_q)$, using (5.11) gives

$$\mathbf{N}_b'\mathbf{N}_b = (\mathbf{I}_b \otimes \mathbf{1}_p')\mathbf{N}_p'\mathbf{N}_q(\mathbf{I}_b \otimes \mathbf{1}_q) = (rpq/b)\mathbf{J}$$

This is the block concurrence matrix of a complete block design for $v = pqb/r$ treatments in b blocks each of size pq. Thus, if (5.11) holds then the block component design D_b is necessarily a complete block design.

Eccleston and Kiefer (1981) show that the information matrices $\mathbf{A}, \mathbf{A}_p$ and $\mathbf{A}_q$ for row–column designs with adjusted orthogonality have a common set of eigenvectors. Further, it can be shown that $r\mathbf{A} = \mathbf{A}_p\mathbf{A}_q$. Hence, the canonical efficiency factors satisfy the two relationships (since $e_{bj} = 1$)

$$e_j = e_{pj} + e_{qj} - 1 \tag{5.12}$$

$$e_j = e_{pj}e_{qj} \tag{5.13}$$

The properties of adjusted orthogonal row–column designs can, therefore, be determined directly from those of its row-component and column-component designs. The harmonic mean efficiency factor E of the row–column design can be expressed as a function of the corresponding efficiency factor E_p and E_q of the component designs. Using (5.12) and (5.13),

$$\frac{1}{e_j} = \frac{1}{e_{pj}e_{qj}} \cdot \frac{e_{pj} + e_{qj}}{1 + e_j}$$

i.e.

$$e_j^{-1} = e_{pj}^{-1} + e_{qj}^{-1} - 1$$

which leads to

$$E = \frac{E_pE_q}{E_p + E_q - E_pE_q} \tag{5.14}$$

It follows, therefore, from (5.13) and (5.14) that if the two component designs are E-, D- or A-optimal then, among the class of adjusted orthogonal designs, the row–column design will also be E-, D- or A-optimal.

A row–column design is said to be *connected* if the rank of the information matrix **A**, given in (5.3), is $v-1$. All contrasts in the treatment parameters will then be estimable. A design will be connected if and only if it has $v-1$ canonical efficiency factors; calculation of these factors provides a simple way of determining whether a given design is connected or not.

Pearce (1975) gives the following example of a disconnected row–column design for $v=4$ treatments with $b=1$ and $p=q=4$:

0	0	2	3
0	0	3	2
2	3	1	1
3	2	1	1

The comparison of treatments 0 and 1 is not estimable; this can easily be verified by showing that the estimability condition of Section 1.4 does not hold for this contrast. Note that, for this design, both the row and column designs D_p and D_q are connected (D_b is degenerate). This example, therefore, shows that the connectedness (or otherwise) of a row–column design cannot necessarily be established from the connectedness (or otherwise) of its component designs. Consider also designs satisfying (5.6). Such designs will be disconnected if $e_{pj}+e_{qj}=e_{bj}$ for some $j(1\leqslant j\leqslant v-1)$ even if all three component designs are connected. As an example, consider the following design for $v=8$, $b=8$, $p=q=2$:

0 1	1 2	2 3	3 4	4 5	5 6	6 7	7 0
3 2	4 3	5 4	6 5	7 6	0 7	1 0	2 1

The information matrices $\mathbf{A}_b, \mathbf{A}_p$ and $\mathbf{A}_q$ are all circulant matrices so that (5.6) holds. The component designs are connected but the row–column design is not since, for one particular value of $j, e_{pj}=e_{qj}=0.5$ and $e_{bj}=1$.

Few general results are available on the connectedness of row–column designs. A result given by Russell (1976) for $b=1$ can be extended to the general case of $b\geqslant 1$ as follows. Suppose that the equally replicated component designs D_b and D_q are both efficiency-balanced with canonical efficiency factors E_b and E_q respectively, i.e.

from (2.11),

$$\mathbf{A}_b = rE_b(\mathbf{I}_v - \mathbf{K}_v) \qquad \mathbf{A}_q = rE_q(\mathbf{I}_v - \mathbf{K}_v)$$

Then, from (5.6), the row–column design will be connected if and only if $E_b - E_q$ is not a canonical efficiency factor of the row component design D_p, i.e. if and only if $rq(E_q - E_b + 1)$ is not an eigenvalue of the concurrence matrix $\mathbf{N}_p\mathbf{N}'_p$; see also Russell (1980).

For designs satisfying the adjusted orthogonality condition of (5.11) the row–column design is connected if and only if the row component and column component designs D_p and D_q are connected. It follows from (5.13) that e_j is not zero if and only if both e_{pj} and e_{qj} are not zero ($1 \leqslant j \leqslant v-1$). This result for $b=1$ was given by Raghavarao and Federer (1975).

Further results on the connectedness of row–column designs for $b=1$ can be found in Butz (1982).

5.5 Row–column and nested row–column designs

It will be convenient in the remainder of this chapter to distinguish between designs having $b=1$ blocks and those with $b>1$ blocks. In the partition of the degrees of freedom given in Table 5.1 it is clear that the block stratum becomes redundant if $b=1$; in the other strata the term 'within blocks' can consequently be deleted. Designs with $b=1$ will, henceforth, be called *row–column designs*, and those for $b>1$ will be called *nested row–column designs*, since the row and column components are nested within blocks.

For row–column designs ($b=1$) the information matrix given in (5.3) then becomes

$$\mathbf{A} = \mathbf{A}_p + \mathbf{A}_q - \mathbf{r}^\delta + (1/pq)\mathbf{r}\mathbf{r}' \tag{5.15}$$

and the vector of adjusted treatment totals given in (5.4) becomes

$$\mathbf{q} = \mathbf{q}_p + \mathbf{q}_q - \mathbf{T} + \bar{y}\mathbf{r} \tag{5.16}$$

If the row and column component designs D_p and D_q have information matrices spanned by a common set of eigenvectors then (5.6) reduces to

$$e_j = e_{pj} + e_{qj} - 1, \tag{5.17}$$

a result given by Pearce (1975).

Row–column designs will be considered in Sections 5.6–5.8 and nested row–column designs in Sections 5.9–5.10. All these designs will have treatments replicated an equal number of times, i.e. $\mathbf{r} = r\mathbf{1}$.

5.6 Latin square designs

If the two component designs D_p and D_q are both randomized block designs, the resulting row–column design is called a Latin square. Such a design necessarily has the number of treatments equal to the number of rows and columns i.e. $v = p = q$. An example of a Latin square with $v = 4$ has been given in Section 5.1. Latin squares are, therefore, row–column designs having each treatment occurring once in each row and once in each column, and are easily constructed for any value of v.

It follows, from the results in Section 5.4, that Latin squares are orthogonal row–column designs, so that estimates of treatment parameters are based on treatment means, no adjustment for rows or columns being necessary. Hence, the contrast $\mathbf{c}'\boldsymbol{\tau} = \sum c_m \tau_m$ in the treatment parameters is estimated by $\sum c_m \bar{y}_m$ with variance $(\sigma^2/r)\sum c_m^2$, where $\bar{y}_m$ is the mth treatment mean.

All the information on treatment contrasts is available from comparisons within rows and columns, since the designs are efficiency-balanced with all canonical efficiency factors equal to unity. The analysis of variance for a Latin square is given in Table 5.2 where $\mathbf{T}, \mathbf{R}$ and $\mathbf{C}$ are vectors of treatment, row and column totals respectively, G is the overall total and $n = v^2$.

More generally, an orthogonal row–column design is obtained if D_p and D_q are both complete block designs. The simplest way of obtaining such designs is by joining together a number of Latin

Table 5.2 *Analysis of variance for a Latin square.*

	d.f.	s.s.
Between rows	$v-1$	$(1/v)\mathbf{R}'\mathbf{R} - G^2/n$
Between columns	$v-1$	$(1/v)\mathbf{C}'\mathbf{C} - G^2/n$
Between treatments	$v-1$	$(1/v)\mathbf{T}'\mathbf{T} - G^2/n$
Residual	$(v-1)(v-2)$	by subtraction
Total	$n-1$	$\mathbf{y}'\mathbf{y} - G^2/n$

squares; the number of rows and columns now being a multiple of the number of treatments.

Example 5.1
An orthogonal row–column design with $v = 5$, $p = 5$ rows and $q = 10$ columns is given by

0	1	2	3	4	0	1	2	3	4
1	3	4	2	0	1	4	3	0	2
2	0	1	4	3	2	0	1	4	3
3	4	0	1	2	3	2	4	1	0
4	2	3	0	1	4	3	0	2	1

obtained by joining together two 5×5 Latin squares. Each treatment now occurs twice in each row and once in each column.

For small values of v, it becomes necessary to use more than one Latin square in order to have sufficient degrees of freedom in the analysis of variance to estimate adequately the experimental error σ^2.

5.7 Row–orthogonal designs

When it is impractical to have a complete replicate of the treatments in each column (or row), incomplete block arrangements will have to be used. This section will be concerned with row–column designs in which each treatment occurs equally often in each row of the design but where some treatments occur less frequently than others in the columns. Thus, the row component design D_p will be a complete block design, so that the number of columns q must be a multiple of the number of treatments v. The column component D_q will usually be a binary incomplete block design. Since the canonical efficiency factors of D_p are all equal to unity, it follows from (5.17) that the canonical efficiency factors of these row–column designs are equal to those of D_q. i.e.

$$e_j = e_{qj} \qquad (j = 1, 2, \ldots, v - 1) \tag{5.18}$$

Hence, choosing an appropriate (e.g. A-optimal) row–column design is equivalent to choosing an appropriate column component design D_q.

From (5.15) and (5.16), the information matrix $\mathbf{A}$ and the vector of adjusted treatment totals $\mathbf{q}$ of these row–column designs are also equal to those of D_q. Estimates of treatment parameters are, therefore, the same as those obtained from a model in which the row parameters have been deleted. These row–column designs will be said to be *row-orthogonal designs*, since rows are orthogonal to treatments. Row effects are only involved in estimating the experimental error in the analysis of variance; with the row sum of squares being of the same form as that given in Table 5.2, with p rather than v rows. Note that rows and columns are interchangeable, so that column-orthogonal designs are obtained by simply interchanging rows and columns in row-orthogonal designs.

The block designs of Chapters 3 and 4 can be set out as the columns of row–column designs. Thus, use of these block designs as the column component D_q will provide classes of row-orthogonal designs. It is necessary, of course, to be able to rearrange the treatments within the blocks so that D_p is a complete block design.

Example 5.2
Consider the balanced incomplete block design for $v = 5$ treatments in 5 blocks of 4 given in Section 3.3. The treatments can be rearranged within blocks to give the row-orthogonal design

0	4	3	2	1
1	0	4	3	2
2	1	0	4	3
3	2	1	0	4

Note that each treatment occurs once in each row.

A row-orthogonal design in which D_q is a balanced incomplete block design is called a *Youden square.* It is always possible to set out any symmetric ($v = q$) balanced incomplete block design as a Youden square.

Example 5.3
As a further example, a Youden square design for $v = 13$ treatments in $p = 4$ rows and $q = 13$ columns, obtained by rearranging the balanced incomplete block design of Example 3.3, is

0	9	8	6	7	2	4	11	3	5	1	12	10
1	3	9	0	10	5	8	6	2	7	12	4	11
2	4	7	3	1	10	0	5	11	12	8	6	9
9	5	6	10	4	8	11	1	7	0	3	2	12

This design might be used, for example, in the preference testing experiment of Section 5.1, where it is felt it was unreasonable for subjects to be expected to discriminate between more than four products (treatments).

Youden square designs are, from (3.5) and (5.18), variance-balanced and optimal under all the optimality criteria given in Section 2.4. Of course, these designs will only exist for certain parameter combinations so that it is necessary to consider designs based on other incomplete block arrangements.

Some group-divisible designs can be rearranged to give designs which are row-orthogonal. The catalogue of PBIB/2 designs given by Clatworthy (1973) sets out the designs in this form if such an arrangement is possible. Not all symmetric group-divisible designs can, however, be so arranged. For example, the design for $v = 9$ treatments in 9 blocks of 3 given in Section 3.5.1 cannot be set out as a row-orthogonal design. Other types of PBIB/2 designs in Clatworthy's catalogue are also set out as row-orthogonal designs where possible. It has already been pointed out in Section 3.1 that such PBIB/2 designs may not be particularly efficient or appropriate to use in practice.

Designs based on cyclical methods of construction provide a flexible class of row-orthogonal designs. A cyclic set with $r = k$ will necessarily have each treatment occurring once in each row, so that all cyclic designs with the number of blocks (i.e. columns) a multiple of the number of treatments are automatically row-orthogonal. The design for $v = 7$ treatments in 7 blocks of 3 given in Section 4.1 illustrates this, although in this case since the design is also a balanced incomplete block design the resulting row–column design is a Youden square. The tables of cyclic designs listed in John et al. (1972), Lamacraft and Hall (1982) and in Tables 4.2–4.4 can be used to provide efficient row-orthogonal designs. The same property holds for the n-cyclic designs of Section 4.5.

Consider now the generalized cyclic designs of Section 4.6 where the $v = mn$ treatments are divided into m groups of n elements using the residue classes, modulo m. The designs are obtained by successive addition of m, modulo v, to be elements of one or more initial blocks. To obtain a symmetric generalized cyclic design, therefore, m initial blocks are generally needed and the design will be row-orthogonal if the elements in the ith position (i.e. ith row) of each of these m blocks belong to different residue classes ($i = 1, 2, \ldots, q$).

Example 5.4

For $v = 8$ treatments in $m = 2$ groups of $n = 4$ the residue classes are

$$S_0 = \{0 \quad 2 \quad 4 \quad 6\} \quad \text{and} \quad S_1 = \{1 \quad 3 \quad 5 \quad 7\}$$

Let one initial block be (0 1 2 4) so that the other initial block must have one element from S_0 and three from S_1. One such block is (0 1 3 7) and, suitably rearranged, leads to the row-orthogonal generalized cyclic design with $p = 4$ and $q = 8$:

0	2	4	6	7	1	3	5
1	3	5	7	0	2	4	6
2	4	6	0	1	3	5	7
4	6	0	2	3	5	7	1

The α-designs of Section 4.8 with $r = k$ can be used to obtain row-orthogonal designs in which the columns are set out in replicate groups. The designs for $v = ks$ treatments in $p = k$ rows and $q = v$ columns are obtained by first constructing an α-design and then reordering the rows within each replicate group so as to ensure that each treatment occurs once in each row.

Example 5.5

From the α-design for $v = 9$ and $k = s = 3$ with generating array

0	0	0
0	1	2
0	2	1

the following row-orthogonal design with 3 rows and 9 columns can be obtained:

Replicate		1			2			3	
Column	1	2	3	4	5	6	7	8	9
Row 1	0	1	2	8	6	7	5	3	4
2	3	4	5	0	1	2	7	8	6
3	6	7	8	4	5	3	0	1	2

The tables of α-designs given in Williams (1975a) provide suitable designs for $r = k = 3$ and $r = k = 4$, while the basic arrays in Table 4.5 provide alternative designs for $r = k = 4$. Designs for $r = k > 4$ have not been tabulated.

It can be seen, therefore, that the block designs of Chapters 3 and 4 provide a means of obtaining a wide choice of row–column designs in which the rows are orthogonal to treatments.

Note that all row-orthogonal designs satisfy the adjusted orthogonality condition of (5.11). For binary designs the condition $\mathbf{N}'_p\mathbf{N}_q = r\mathbf{J}$ means that each column of the design has r treatments in common with each row, which is a condition clearly satisfied by row-orthogonal designs since $r = p$. Thus, (5.18) could also be obtained from (5.13) as $e_{pj} = 1$ for all j.

5.8 Row–column α-designs

Suppose now that neither D_p nor D_q is a complete block design, which is a necessary feature of any row–column design in which neither p nor q is a multiple of v. A large number of designs can again be obtained using cyclical methods of construction, and a flexible class of designs based on the α-designs of Patterson and Williams (1976a) will be described in this section. Other small classes of designs and a few *ad hoc* designs are also available. Pearce (1963), for instance, gives the following variance-balanced design for $v = 4$, $p = q = 6$ and $r = 9$:

0	3	2	1	0	3
1	1	3	0	2	2
3	2	1	2	3	0
2	0	3	0	1	1
2	2	1	3	0	0
1	3	0	3	1	2

Both D_p and D_q are (non-binary) variance-balanced designs, although this requirement is not necessary for the row–column design to be variance-balanced. Kshirsagar (1957) gives the following variance-balanced design for $v = 9$, $p = q = 6$ and $r = 4$ in which both D_p and D_q are (Latin square type) PBIB/2 designs:

0	7	1	3	8	5
1	3	7	6	5	2
2	4	6	1	3	8
3	8	0	7	2	4
4	5	2	0	6	7
5	6	8	4	0	1

Freeman (1957) gives a number of designs in which D_p and D_q are both PBIB/2 designs, although the resulting row–column designs are not variance-balanced. Shrikhande (1951) and Agrawal (1966) give some designs with two distinct canonical efficiency factors. Raghavarao and Shah (1980) give a class of designs for $v = 2p$ treatments in $p \geqslant 4$ rows and $q = 2(p-1)$ columns which satisfy the adjusted orthogonality condition of (5.11), but these designs can be shown to belong to the wider class of row–column α-designs described below. Russell (1980) gives a simple algorithm to construct (M,S)-optimal designs when D_p is a balanced incomplete block design and when $\binom{q}{t}$ divides v, where t/q is the fractional part of r/q; trial-and-error methods were used to obtain designs in the four cases when $\binom{q}{t}$ does not divide v. When $t = 0$ the optimal design is a Youden square.

John and Eccleston (1986) give a class of row–column designs based on the α-designs of Section 4.8. Consider again the row-orthogonal design with $v = q = 9$ and $k = s = p = 3$ given in Example 5.5. Suppose the last replicate (i.e. the last three columns) is omitted. The resulting design is equally replicated ($r = 2$) with $p = 3$ rows and $q = 6$ columns. Further, it can be seen to satisfy the adjusted orthogonality condition of (5.11) since each column has two treatments in common with every row.

In general, the α-designs for $v = ks$ treatments can be used to give a class of row–column designs with $p = k$ rows and $q = rs$ columns having the property of adjusted orthogonality. It will be assumed that $q < v$ so that D_p will be a binary incomplete block design, although the method of constructing the designs can be readily extended to the case where $q > v$. Let the v treatments be divided into k classes of s elements such that the ith class is

$$T_i = \{is, is+1, \ldots, (i+1)s-1\} \qquad (i = 0, 1, \ldots, k-1) \quad (5.19)$$

The designs are obtained by first constructing an α-design using the method given in Section 4.8.1. Note that at this stage the ith row contains elements only from class T_i $(i=0,1,\ldots,k-1)$. The rows within each replicate are then reordered to ensure that each row of the resulting row–column design contains at most one element from each of the classes T_i $(i=0,1,\ldots,k-1)$.

Example 5.6

Consider the α-design for $v=12$, $k=4$ and $r=s=3$ given in Example 4.4. The rows within each replicate can now be reordered to give the following design with $p=4$ rows and $q=9$ columns:

0	1	2	3	4	5	7	8	6
3	4	5	8	6	7	10	11	9
6	7	8	10	11	9	0	1	2
9	10	11	0	1	2	5	3	4

It can be verified that each column has three treatments in common with every row, i.e. $\mathbf{N}_p'\mathbf{N}_q = 3\mathbf{J}$.

In view of the properties of adjusted orthogonal designs, the efficiency factors of these row–column α-designs can be obtained directly from those of the row component design D_p and column component design D_q. Since D_q is an α-design the tables of Williams (1975a) or the basic arrays in Table 4.5 can be used to obtain suitable column component designs. The row component D_p is given by considering the way in which the classes T_i are represented within each replicate of the final design. For the design of Example 5.6 the arrangement is

$$\begin{array}{ccc} T_0 & T_1 & T_2 \\ T_1 & T_2 & T_3 \\ T_2 & T_3 & T_0 \\ T_3 & T_0 & T_1 \end{array}$$

which is an arrangement of four objects (T_0, T_1, T_2 and T_3) in four groups (or blocks) of 3. In general, the arrangement of the classes in this manner produces a symmetric incomplete block design with k objects in k blocks of r. Now if this symmetric design, D_p^* say, with parameters $v^*=b^*=k$ and $k^*=r^*=r$ has incidence matrix $\mathbf{N}_p^*$, then the incidence matrix of D_p is given by

$$\mathbf{N}_p = \mathbf{N}_p^* \otimes \mathbf{1}_s \qquad (5.20)$$

Assuming D_p^* to be connected, the $v-1$ canonical efficiency factors of D_p are then given by the canonical efficiency factors of D_p^* together with $k(s-1)$ factors equal to 1 corresponding to contrasts between treatments within the T_i-classes. Hence, if E_p^* and E_p are the average efficiency factors of D_p^* and D_p respectively, then

$$E_p = \frac{ks-1}{k(s-1)+(k-1)E_p^{*-1}} \tag{5.21}$$

If D_p^* is disconnected then so are D_p and the row–column design. The choice of a good row component design, therefore, becomes one of choosing a good D_p^* design. The designs of Chapters 3 and 4 can be used for this purpose, where treatment label i is replaced by T_i.

Eccleston and Russell (1980) have shown that if both D_p and D_q are (M, S)-optimal then the adjusted orthogonal row–column design will be (M, S)-optimal among *all* row–column designs of the same size. One particular case where such designs can often be constructed is when $r=k-1$ and $k \leqslant s$. To achieve this an unreduced balanced incomplete block design is used for D_p^*, so that D_p is a singular group-divisible design with concurrences $\lambda_1 = k-1$ and $\lambda_2 = k-2$, and an $\alpha(0,1)$-design, assuming one exists, is used for D_q.

Example 5.7
The following 3×8 design for $v=12$, $k=3$, $s=4$ is (M, S)-optimal:

0	1	2	3	10	11	8	9
4	5	6	7	0	1	2	3
8	9	10	11	5	6	7	4

When $4 \leqslant k \leqslant 10$, $k \leqslant s$ and $r=2$, 3 or 4, a large class of efficient designs is readily provided from the use of the basic arrays of Table 4.5 as the source of the column component designs and with cyclic designs for k treatments having initial block (0 1 3 x) as the source of row component designs, where

$$x = \begin{cases} 2 & \text{if } k = 4, 5, 6 \\ 5 & \text{if } k = 7, 10 \\ 6 & \text{if } k = 8, 9 \end{cases}$$

This will produce designs for $r=4$. Deleting the last replicate (i.e. the last s columns) will give designs for $r=3$, while deleting the last two replicates gives designs for $r=2$. The row component designs will be efficient, as the cyclic designs for k treatments with initial blocks (0 1) or (0 1 3) or (0 1 3 x) are, with two exceptions, A-optimal within the class of designs whose concurrences differ by no more than one. The exceptions are the designs with $k = 9$, $r = 3$ and $k = 10$, $r = 3$, but even here there is little loss in efficiency factors.

Example 5.8

The following α array for $k = 5$ and $r = 4$ is obtained from the basic array with $s = k = 6$ in Table 4.5:

0	0	0	0
0	1	5	4
0	3	2	5
0	2	3	1
0	4	1	2

Using the cyclic design with initial block (0 1 3 2) for D_p^* leads to the row–column design for $v = 30$ in 5 rows and 24 columns:

0	1	2	3	4	5	7	8	9	10	11	6	21	22	23	18	19	20	17	12	13	14	15	16
6	7	8	9	10	11	15	16	17	12	13	14	25	26	27	28	29	24	19	20	21	22	23	18
12	13	14	15	16	17	20	21	22	23	18	9	0	1	2	3	4	5	26	27	28	29	24	25
18	19	20	21	22	23	28	29	24	25	26	27	11	6	7	8	9	10	0	1	2	3	4	5
24	25	26	27	28	29	0	1	2	3	4	5	14	15	16	17	12	13	10	11	6	7	8	9

Deleting the last 6 and 12 columns gives efficient row–column α-designs for $p = 5$, $q = 18$ and $p = 5$, $q = 12$ respectively.

The above method of constructing adjusted orthogonal designs can be generalized to give designs with $p = tk$ rows, where $t > 1$, if t-resolvable α-designs are used. A design is *t-resolvable* if the blocks can be set out in groups such that each treatment is replicated t times within a group. An intermediate array α^* is obtained from a $tk \times r$ array α as in Section 4.8.1. Now, however, s is added to each element in rows $t+1, t+2, \ldots, 2t$; $2s$ is added to each element in rows $2t+1$, $2t+2, \ldots, 3t$; and so on. Within each replicate group rows are again reordered according to a design D_p^* for k objects in tk blocks of size r. The choice of an appropriate column component design is no longer straightforward, as efficient t-resolvable α-designs have not been tabulated, although some of the generalized cyclic designs listed in Hall and Jarrett (1981) can be shown to be t-resolvable α-designs.

5.9 Lattice squares

The remainder of this chapter is concerned with nested row–column designs, i.e. with designs in which v treatments are set out in $b > 1$ blocks of $k = pq$ units, where the treatments within each block are then arranged in a row–column design of p rows and q columns. A fully orthogonal nested row–column design is given by taking b replicates of a Latin square. Other row–column designs can also be replicated to provide further nested designs. In particular, if the row-orthogonal designs of Section 5.7 are replicated then the properties of these designs are again given by those of the column component design D_q. The size of such designs will necessarily be large, and alternative designs using very much smaller values of p and q can be obtained.

Lattice squares, introduced by Yates (1940a), are variance-balanced nested row–column designs for $v = s^2$ treatments set out in $b = r$ blocks with $p = q = s$, where $r = \frac{1}{2}(s + 1)$ for s odd, and $r = s + 1$ for s even. In Section 3.4.2, balanced lattice designs were defined for $v = s^2$ treatments using complete sets of mutually orthogonal Latin squares. The s^2 treatments were set out in an $s \times s$ array and $s + 1$ replicates of s blocks each obtained by superimposing rows, columns and each of the $s - 1$ orthogonal Latin squares onto the array. Such designs exist for any s which is a prime or a power of a prime. In a lattice square the rows and columns correspond to these replicates. They are arranged in such a way that for s odd, half of the replicates constitute the rows and the other half the columns, whereas for s even, each replicate is used both for rows and for columns.

Example 5.9
For $s = 4$ the five replicates (blocks) are:

Replicate 1				2				3				4				5			
0	1	2	3	0	4	8	12	0	5	10	15	0	6	11	13	0	7	9	14
4	5	6	7	5	1	13	9	11	14	1	4	7	1	12	10	1	6	8	15
8	9	10	11	10	14	2	6	13	8	7	2	9	15	2	4	2	5	11	12
12	13	14	15	15	11	7	3	6	3	12	9	14	8	5	3	3	4	10	13

Note that the elements in each row of a replicate have been arranged in such a way that the columns of the first replicate correspond to the rows of the second replicate, the columns of the second to the rows of the third, and so on. Hence, both the row and column component designs D_p and D_q are balanced lattice designs and the resulting nested row–column design is a lattice square with $v = 16$, $b = r = 5$ and $p = q = 4$.

The lattice square for $s = 3$ given in Section 5.1 has the rows of the two squares corresponding to two of the $s + 1 = 4$ replicates of a balanced lattice design, while the columns correspond to the other two replicates. Hence, D_p and D_q together give a balanced lattice design. In all cases, the block component design D_b is a randomized block design.

The lattice square with $r = s + 1$ is clearly variance-balanced, from (5.6), with $\mathbf{N}_p\mathbf{N}_p' + \mathbf{N}_q\mathbf{N}_q' = 2(s\mathbf{I} + \mathbf{J})$. For $r = \frac{1}{2}(s + 1)$, both D_p and D_q are (Latin-square type) PBIB/2 designs based on the same association scheme with $\lambda_1 = 1$, $\lambda_2 = 0$ for D_p and $\lambda_1 = 0$, $\lambda_2 = 1$ for D_q. Hence, $\mathbf{N}_p\mathbf{N}_p' + \mathbf{N}_q\mathbf{N}_q' = s\mathbf{I} + \mathbf{J}$ so that, using (5.3), it is seen that these lattice squares are also variance-balanced. All the canonical efficiency factors of a lattice square design are equal to

$$E = (s - 1)/(s + 1) \qquad (5.22)$$

so that the information matrix $\mathbf{A}$ is

$$\mathbf{A} = rE(\mathbf{I}_v - \mathbf{K}_v) \qquad (5.23)$$

Thus the analysis of a lattice square follows that of a balanced lattice design, with the information matrix given by (5.23) and the vector of adjusted treatment totals calculated using (5.16).

5.10 Other nested row–column designs

Variance-balanced nested row–column designs whose block component designs D_b are binary incomplete block designs, rather than randomized block designs, have been given by Preece (1967), Singh and Dey (1979), Street (1981), Agrawal and Prasad (1982a) and Ipinyomi and John (1985). The information matrix of these designs is again given by (5.23) but with

$$E = \frac{v(p - 1)(q - 1)}{pq(v - 1)} \qquad (5.24)$$

Note that when $v = s^2$ and $p = q = s$, this value of E is equal to that of a lattice design given in (5.22).

One method of construction, given by Singh and Dey (1979), is to take a balanced incomplete block design for v treatments in b blocks of $k = s^2$ treatments per block and then to set out the s^2 treatments in each of these blocks in a lattice square of $s + 1$ blocks. However, the resulting balanced row–column design will invariably require a very large number of blocks, namely $b(s + 1)$ blocks.

Other balanced designs can be obtained by cyclically developing one or more initial row–column designs. Agrawal and Prasad (1982a) give four series of designs using this method, which require that $v = tpq + 1$ for some $t \geqslant 1$ and that v is a prime power. Ipinyomi and John (1985) have considered the general problem of obtaining nested row–column designs based on cyclic and generalized cyclic methods of construction, and have tabulated a number of designs for $v \leqslant 15$, nine of which are balanced.

As was the case with incomplete block designs, variance-balanced nested row–column designs will only exist for a limited number of parameter combinations. Most designs involve a large number of blocks and have, therefore, limited practical use. Agrawal and Prasad (1982b) give two series of designs, called group-divisible and rectangular nested row–column designs respectively, for which the block, row and column components are all balanced incomplete blocks. Again these designs, when they do exist, involve a relatively large number of blocks.

A flexible family of nested row–column designs can be obtained using cyclical methods of construction. Jarrett and Hall (1982) suggested the use of the generalized cyclic designs of Section 4.6, while Ipinyomi and John (1985) have provided a table of efficient designs for $5 \leqslant v \leqslant 15$, $p \leqslant 3$, $q \leqslant 5$ and $b \leqslant v$.

Example 5.10
Consider the following design for 7 treatments in 7 blocks of 4, with each block set out in 2 rows and 2 columns:

0	2		1	3		2	4		3	5		4	6		5	0		6	1
4	1		5	2		6	3		0	4		1	5		2	6		3	0

The blocks of 2×2 squares have been generated by a cyclical development of the first block.

If the three component designs D_b, D_p and D_q are cyclic designs, as they are in Example 5.10, then their information matrices will be circulant matrices. The canonical efficiency factors of these cyclic nested row–column designs can, therefore, be obtained directly from the information matrix $\mathbf{A}$ of (5.3), or from the component designs using (5.6), or by calculating

$$e_u = (1/r) \sum_{h=0}^{v-1} (a_{ph} + a_{qh} - a_{bh}) \cos(2\pi hu/v) \qquad (u = 1, 2, \ldots, v-1) \tag{5.25}$$

where a_{ph}, a_{qh} and a_{bh} are the elements in the first row and $(h+1)$th column of the information matrices $\mathbf{A}_p$, $\mathbf{A}_q$ and $\mathbf{A}_b$ respectively. This formula follows directly from (4.2) and (5.3).

Each initial block of D_b will usually generate a full arrangement of $b = v$ blocks, although partial arrangements with $b < v$ can also be obtained if v and k have a common divisor; see Section 4.2. In order to be able to use (5.25) for designs with $b < v$ it is necessary, however, that not only should D_b be a partial set but that both D_p and D_q should still be cyclic designs, consisting of full sets, partial sets or multiples of partial sets. For example, with $v = 12$ and $k = 9$ the initial block (0 1 2 4 5 6 8 9 10) will generate a partial cyclic set with $b = 4$ and $r = 3$. Consider the following two ways in which this initial block can be set out in an array with $p = 3$ rows and $q = 3$ columns:

0	1	2		0	1	2
5	6	4		5	4	6
10	8	9		8	9	10

Cyclic development of the first array will produce a cyclic row–column design for $b = 4$ blocks, whereas the second array will not; it can be checked that with four blocks the column component D_q is not a cyclic design. The second array will, of course, produce a cyclic row–column design with $b = 12$ blocks.

Alternatively, when $b < v$ generalized cyclic methods of construction can be used.

Example 5.11

The component designs of the following nested row–column design for $v = 8$, $b = 4$, $r = 3$, $p = 2$ and $q = 3$ are all generalized cyclic designs with increment number $m = 2$:

0	1	4	2	3	6	4	5	0	6	7	2
2	5	3	4	7	5	6	1	7	0	3	1

When $b < v$, Ipinyomi and John (1985) show that for some values of v, p, q and r the best cyclic row–column designs are based on partial cyclic sets ($m = 1$), but for other parameter values they are obtained from generalized cyclic designs ($m > 1$).

CHAPTER 6

Factorial experiments: single and fractional replication

6.1 Introduction

Previous chapters have been concerned with experiments in which a single set of treatments was applied to the plots in a block design or in a row–column design. Suppose now that there are two different treatments, say temperature and pressure, and that their effect on the response of some chemical process is to be studied. One approach would be to carry out two separate experiments. One would be concerned with the effect produced by varying temperature levels, with pressure kept at some constant value. The other experiment would look at the effect produced by varying pressure levels with temperature now held constant.

An alternative approach would be to study the effect of temperature and pressure simultaneously. An experiment could be carried out in which different combinations of the two treatments, temperature and pressure, are used and their effects assessed. Such an experiment is an example of a *factorial experiment*. The effect of temperature can now be obtained either at each different level of pressure or averaged over the levels of pressure. In a similar way, the effects of pressure can be examined. Additionally, a factorial experiment provides information on how the two treatments *interact* with each other. It may be, for instance, that the effect of temperature is different at different levels of pressure. Such information is not available when carrying out separate experiments. Only by changing the experimental conditions in this way is it possible to examine not only the effects of temperature and pressure separately but also the way they interact.

As a further example, consider an experiment to compare the weight increases of animals kept on different diets using animals of different breeds fed by different methods. Using a factorial experiment it will be possible to determine the best diets, breeds and methods. Also it will be possible to study whether, for instance, the best method of feeding varies from diet to diet or whether the optimum method and diet depend upon the breed of the animal.

Each basic treatment, such as temperature or diet, will be called a *factor*. The number of possible forms of a factor will be called the number of *levels* of the factor. For example, if there are three diets then the factor 'diets' will be at three levels. A particular combination of factors determines a treatment or *treatment combination*. A *symmetrical* factorial experiment or design will have all factors at the same number of levels, otherwise it will be *asymmetrical*. The effect of a factor averaged over the levels of all the other factors is called the *main effect* of that factor. The way in which the effects of factors change at different levels of other factors represents the *interaction* between factors.

Treatment combinations can be set out in any of the designs given in earlier chapters. However, attention will be restricted primarily to factorial experiments in block designs; the use of row–column designs will be discussed in Chapter 7. Again the allocation of treatment combinations to plots should be aimed at maximizing the amount of information on treatments from comparisons made within blocks. With complete block designs all treatment contrasts are estimated entirely within blocks, but constraints on block size often make the use of such designs impractical. If the number of treatment combinations is not too large then it is often possible to set out the experiment in incomplete block designs in which some information is available within blocks on all factorial effects (main effects and interactions); such effects are said to be *partially confounded* with blocks. However, factorial experiments with many factors, or with factors at many levels, involve large numbers of treatment combinations. The use of designs which require a number of replicates of each treatment combination then becomes impractical. For instance, an experiment with 4 factors each at 3 levels involves 81 treatment combinations so that a block design with only two replicates requires taking 162 observations.

To overcome this problem, designs using just a single replicate are frequently employed. Information on all or part of some of the factorial effects will consequently no longer be available from comparisons within blocks; these effects, or some components of them, will be said to be *totally confounded* with blocks. A further problem is that, since there are no treatment replications, an estimate of error has to be obtained by assuming that certain effects, usually the high-order interactions, are negligible. Even single replicate designs are sometimes too large to use in practice so that fractional designs have to be used; these are discussed in Section 6.10.

The principle of confounding will now be illustrated by considering in detail a factorial experiment with two factors F_1 and F_2 each at three levels, the 3^2 factorial experiment. Before doing so, however, main effects and interaction in the two-factor experiment will be defined.

More generally, let the two factors F_1 and F_2 be at m_1 and m_2 levels respectively. A treatment combination will be denoted by the 2-tuple a_1a_2, where a_i represents a level of factor F_i ($a_i = 0, 1, \ldots, m_i - 1$; $i = 1, 2$). For the 3^2 experiment the 9 treatment combinations are, therefore, given by

$$00 \quad 01 \quad 02 \quad 10 \quad 11 \quad 12 \quad 20 \quad 21 \quad 22$$

Suppose that all m_1m_2 treatment combinations are accommodated in a single block, and let y_{ij} be the observation obtained from applying treatment combination ij. Often it will be more convenient to represent y_{ij} simply by ij, letting ij stand for both the treatment combination and its corresponding observation. Let

$$\bar{y}_{i.} = \sum_j y_{ij}/m_2, \qquad \bar{y}_{.j} = \sum_i y_{ij}/m_1, \qquad \bar{y} = \sum_i \sum_j y_{ij}/m_1m_2$$

The effect of factor F_1 at level i is measured by

$$F_{1i} = \bar{y}_{i.} - \bar{y} \qquad (i = 0, 1, \ldots, m_1 - 1) \tag{6.1}$$

and is the comparison of the mean of the observations having factor F_1 at level i with the overall mean. Alternatively, it can be regarded as the deviation $y_{ij} - \bar{y}_{.j}$ averaged over the levels of factor F_2. The effects $F_{10}, F_{11}, \ldots$ are (treatment) contrasts in the observations and measure the *main effect* of factor F_1. It is a measure of the extent to which levels of F_1 differ when averaged over the levels of factor F_2. Note that these contrasts are not linearly independent since they sum to zero. There are in fact $(m_1 - 1)$ linearly independent contrasts, so that the main effect of F_1 is based on $(m_1 - 1)$ degrees of freedom.

Similarly, the main effect of factor F_2, with $(m_2 - 1)$ degrees of freedom, is defined in terms of

$$F_{2j} = \bar{y}_{.j} - \bar{y} \qquad (j = 0, 1, \ldots, m_2 - 1) \tag{6.2}$$

and again measures the extent to which levels of factor F_2 differ when averaged over the levels of factor F_1.

Suppose now that the effect of factor F_1 at level i is the same when measured at each level of factor F_2, i.e.

$$F_{1i} = y_{ij} - \bar{y}_{.j}, \qquad \text{for all } j \tag{6.3}$$

If (6.3) is true for all levels of factor F_1 then the two factors F_1 and F_2 are said to be *additive*, since the difference between any two observations y_{ij} and $y_{i'j'}$ is then given by the sum of differences in their (main) effects, i.e.

$$\begin{aligned} y_{ij} - y_{i'j'} &= (\bar{y}_{i.} - \bar{y}_{i'.}) + (\bar{y}_{.j} - \bar{y}_{.j'}) \\ &= (F_{1i} - F_{1i'}) + (F_{2j} - F_{2j'}) \end{aligned}$$

If two factors are not additive then they are said to *interact*. The interaction between the two factors F_1 and F_2, the F_1F_2 interaction, is then measured by the set of contrasts $(y_{ij} - \bar{y}_{.j}) - F_{1i}$, i.e.

$$(F_1F_2)_{ij} = y_{ij} - \bar{y}_{i.} - \bar{y}_{.j} + \bar{y} \qquad \text{for all } i, j \tag{6.5}$$

Factors F_1 and F_2 are interchangeable in (6.3) so that the contrasts (6.5) are also given by $(y_{ij} - \bar{y}_{i.}) - F_{2j}$. The interaction contrasts (6.5) are not linearly independent and, since $\sum_i (F_1F_2)_{ij} = \sum_j (F_1F_2)_{ij} = 0$, are based on $(m_1 - 1)(m_2 - 1)$ degrees of freedom.

Finally, the following identity will be useful in the examples which follow:

$$y_{ij} = F_{1i} + F_{2j} + (F_1F_2)_{ij} + \bar{y} \tag{6.6}$$

Returning now to the 3^2 experiment, if all 9 treatment combinations can be accommodated in a single block then both main effects and the interaction can be estimated within blocks. Consider, however, the following single replicate design for the 3^2 experiment in 3 blocks of 3:

(00 11 22)
(01 12 20)
(02 10 21)

Certain treatment contrasts are no longer estimable within blocks since the design is disconnected. For instance, there are two linearly independent treatment contrasts involving block totals which are not estimable. Thus, letting B_1, B_2 and B_3 represent the three block totals, the block contrast $B_1 - B_3$ corresponds to a contrast in the treatment combinations, namely $(00+11+22)-(02+10+21)$. This treatment contrast is said to be totally confounded with the block contrast. Another treatment contrast is confounded with $B_1 - 2B_2 + B_3$, a block contrast orthogonal to $B_1 - B_3$. Hence, two independent treatment contrasts are totally confounded with blocks. It now remains to show that the two confounded treatment contrasts correspond, in this design, to two degrees of freedom of the F_1F_2 interaction.

It is clear from the design that main effects are estimable. Every level of a factor occurs once in each block so that main effect contrasts can be estimated independently of the block totals. For instance,

$$F_{10} - F_{11} = \tfrac{1}{3}[(00-11)+(01-12)+(02-10)]$$

involving three differences, each of which is estimable within blocks. It follows that components of the F_1F_2 interaction must be confounded with blocks. This can be demonstrated more rigorously by showing that the block contrasts can be expressed as linear combinations of the interaction contrasts $(F_1F_2)_{ij}$ given in (6.5). Using (6.6) it can be verified that, writing f_{ij} for $(F_1F_2)_{ij}$,

$$B_1 - B_3 = (f_{00}+f_{11}+f_{22}) - (f_{02}+f_{10}+f_{21})$$

and

$$B_1 - 2B_2 + B_3 = (f_{00}+f_{11}+f_{22}) - 2(f_{01}+f_{12}+f_{20}) + (f_{02}+f_{10}+f_{21})$$

It can also be noted that the three treatment combinations in the first block of the design satisfy the relationship $a_1 + 2a_2 = 0 \pmod 3$, where again a_i denotes a level of the ith factor $(i = 1,2)$. Further, the treatments in the second block satisfy $a_1 + 2a_2 = 1 \pmod 3$ and those in the third block $a_1 + 2a_2 = 2 \pmod 3$. The two degree of freedom component of the F_1F_2 interaction involving contrasts in the three sets of treatment combinations satisfying the equations

$$a_1 + 2a_2 = 0, 1, 2 \qquad (\text{mod } 3)$$

will be denoted by $F_1F_2^2$; the powers attached to the letters are the coefficients in the linear combination of a_1 and a_2.

The remaining two degrees of freedom of the F_1F_2 interaction, which are necessarily estimable within blocks, correspond to the F_1F_2 component of the interaction, i.e. to contrasts in the sets of treatment combinations satisfying the equations

$$a_1 + a_2 = 0, 1, 2 \qquad (\text{mod } 3)$$

The two components F_1F_2 and $F_1F_2^2$ represent an orthogonal decomposition of the F_1F_2 interaction. The four linearly independent contrasts defined by these components span the interaction space given by the contrasts (6.5).

Not all single replicate designs will lead to components of certain main effects and interactions being totally confounded with blocks. Consider, for instance, the following alternative design for the 3^2 experiment in 3 blocks of 3:

(00 01 22)
(12 20 21)
(02 10 11)

It follows from (6.6) that, for instance

$$B_1 - B_3 = (2F_{10} + F_{12} + f_{00} + f_{01} + f_{22}) \\ - (F_{10} + 2F_{11} + f_{02} + f_{10} + f_{11})$$

so that part of the main effect of F_1 and part of the F_1F_2 interaction are confounded with blocks.

Attention is restricted in this chapter to designs typified by the first example, in which each treatment contrast confounded with blocks belongs entirely within a particular main effect or interaction.

6.2 Treatment structure

Consider again a factorial experiment with two factors F_1 and F_2 at m_1 and m_2 levels respectively. If these treatment combinations are set out in a block design then the appropriate model is given by (1.1), namely

$$y_{ikl} = \mu + \tau_i + \beta_k + \varepsilon_{ikl}$$

where, in particular, τ_i is the effect of the ith treatment combination. It will be more convenient in a two-factor experiment to write the treatment effect as τ_{ij}, with the two suffices identifying the levels of the two factors. Thus, τ_{ij} is the effect of the treatment combination having factor F_1 at the ith level and F_2 at the jth level. Now consider the following identity

$$\tau_{ij} = \bar{\tau} + (\bar{\tau}_{i.} - \bar{\tau}) + (\bar{\tau}_{.j} - \bar{\tau}) + (\tau_{ij} - \bar{\tau}_{i.} - \bar{\tau}_{.j} + \bar{\tau}) \qquad (6.7)$$

where $\bar{\tau}_{i.}$ and $\bar{\tau}_{.j}$ are means averaged over the levels of F_2 and F_1 respectively, and $\bar{\tau}$ is the overall mean of the τ_{ij}. When compared with (6.1) and (6.2), it is clear that the second and third terms on the right side of (6.7) represent main effects of F_1 and F_2 respectively, while from a comparison with (6.5) the final term represents the F_1F_2 interaction. Hence, (6.7) gives a partition of the $m_1m_2 - 1$ linearly independent treatment contrasts $(\tau_{ij} - \bar{\tau})$ into three (orthogonal) components representing the main effects and interaction based on $m_1 - 1$, $m_2 - 1$ and $(m_1 - 1)(m_2 - 1)$ degrees of freedom respectively. This partition defines the treatment structure of the factorial experiment.

In matrix notation, the set of equations (6.7) can be written

$$\boldsymbol{\tau} = \mathbf{C}_{00}\boldsymbol{\tau} + \mathbf{C}_{10}\boldsymbol{\tau} + \mathbf{C}_{01}\boldsymbol{\tau} + \mathbf{C}_{11}\boldsymbol{\tau} \tag{6.8}$$

where

$$\begin{aligned}
\mathbf{C}_{00} &= \mathbf{K}_{m_1} \otimes \mathbf{K}_{m_2} \\
\mathbf{C}_{10} &= (\mathbf{I}_{m_1} - \mathbf{K}_{m_1}) \otimes \mathbf{K}_{m_2} \\
\mathbf{C}_{01} &= \mathbf{K}_{m_1} \otimes (\mathbf{I}_{m_2} - \mathbf{K}_{m_2}) \\
\mathbf{C}_{11} &= (\mathbf{I}_{m_1} - \mathbf{K}_{m_1}) \otimes (\mathbf{I}_{m_2} - \mathbf{K}_{m_2})
\end{aligned}$$

where $\mathbf{K}_{m_j}$ is an $m_j \times m_j$ matrix with every element equal to m_j^{-1}. Premultiplying a vector by the matrix $\mathbf{K}$ results in each element of the vector being replaced by its mean, whereas premultiplying by $\mathbf{I} - \mathbf{K}$ results in the mean being subtracted from each element in the vector. Hence, premultiplying $\boldsymbol{\tau}$ by $\mathbf{C}_{00}$ can be seen as an operation which first averages over the levels of F_2 and then over the levels of F_1 to give a vector with each element equal to $\bar{\tau}$. Premultiplying $\boldsymbol{\tau}$ by $\mathbf{C}_{10}$ first averages over the levels of F_2 and then subtracts the overall mean from each resulting mean. Similarly, premultiplying $\boldsymbol{\tau}$ by $\mathbf{C}_{01}$ results in the overall mean being subtracted from means obtained by averaging over the levels F_1. The product $\mathbf{C}_{11}\boldsymbol{\tau}$ can best be understood by looking at the two parts of the operation. First, premultiplying by $\mathbf{I} \otimes (\mathbf{I} - \mathbf{K})$ produces a vector of differences $\tau_{ij} - \bar{\tau}_{i.}$ where averaging is over the levels of F_2. Premultiplying by $\mathbf{K} \otimes (\mathbf{I} - \mathbf{K})$ now results in these differences being averaged over the levels of F_1 to give differences $\bar{\tau}_{.j} - \bar{\tau}$. Putting the two operations together gives a vector whose elements are of the form

$$(\tau_{ij} - \bar{\tau}_{i.}) - (\bar{\tau}_{.j} - \bar{\tau}) = \tau_{ij} - \bar{\tau}_{i.} - \bar{\tau}_{.j} + \bar{\tau}$$

which is the required contrast.

A concise representation of the **C** matrices in (6.8) is given by

$$\mathbf{C}_{x_1x_2} = \mathbf{C}_{x_1} \otimes \mathbf{C}_{x_2}$$

where

$$\mathbf{C}_{x_j} = \begin{cases} \mathbf{K}_{m_j} \quad, & x_j = 0 \\ \mathbf{I}_{m_j} - \mathbf{K}_{m_j}, & x_j = 1 \end{cases} \tag{6.9}$$

For three factors F_1, F_2 and F_3 at m_1, m_2 and m_3 levels respectively, the vector of treatment parameters $\boldsymbol{\tau}$ can be partitioned into 8 components given by

$$\boldsymbol{\tau} = \sum \mathbf{C}_{x_1x_2x_3}\boldsymbol{\tau} \qquad (x_i = 0, 1, i = 1, 2, 3)$$

where

$$\mathbf{C}_{x_1x_2x_3} = \mathbf{C}_{x_1} \otimes \mathbf{C}_{x_2} \otimes \mathbf{C}_{x_3}$$

and where $\mathbf{C}_{x_j}$ is given by (6.9). The terms in the summation correspond to, respectively, the mean effect, the main effects of F_1, F_2 and F_3, the two-factor interactions F_1F_2, F_1F_3 and F_2F_3 and, additionally, the three-factor interaction $F_1F_2F_3$. This last interaction measures the extent to which the F_1F_2 interaction varies at different levels of F_3 or, equivalently, the F_1F_3 interaction at different levels of F_2 or the F_2F_3 interaction at different levels of F_1. A negligible $F_1F_2F_3$ interaction means that the interaction of any two of the factors is the same at each level of the third factor.

For an n-factor experiment with factor F_i at m_i levels ($i = 1, 2, \ldots, n$) the treatment structure is given by

$$\boldsymbol{\tau} = \sum_x \mathbf{C}_x \boldsymbol{\tau} \tag{6.10}$$

where the summation is over all binary numbers $x = (x_1 x_2 \ldots x_n)$ and

$$\mathbf{C}_x = \mathbf{C}_{x_1} \otimes \mathbf{C}_{x_2} \otimes \cdots \otimes \mathbf{C}_{x_n} = \bigotimes_{j=1}^{n} \mathbf{C}_{x_j} \tag{6.11}$$

and where $\mathbf{C}_{x_j}$ is given by (6.9). A *generalized interaction* (i.e. main effect or interaction) is represented by $\mathbf{C}_x\boldsymbol{\tau}$, where factor F_i is present in the interaction if $x_i = 1$ ($i = 1, 2, \ldots, n$). This generalized interaction will be denoted by $F^x = F_1^{x_1} F_2^{x_2} \ldots F_n^{x_n}$. For example, in a five-factor experiment the $F_1F_2F_4$ interaction is represented by the set of contrasts $\mathbf{C}_{11010}\boldsymbol{\tau}$.

These $\mathbf{C}_x$ matrices form a *complete binary set* in that they are symmetric ($\mathbf{C}_x = \mathbf{C}'_x$), idempotent ($\mathbf{C}_x^2 = \mathbf{C}_x$), orthogonal ($\mathbf{C}_x\mathbf{C}_y = \mathbf{0}$,

$x \neq y$) and satisfy $\sum \mathbf{C}_x = \mathbf{I}$. These results follow as both $\mathbf{K}$ and $(\mathbf{I}-\mathbf{K})$ are symmetric idempotent matrices, as $\mathbf{K}(\mathbf{I}-\mathbf{K}) = \mathbf{0}$, and by noting that

$$\sum_x \mathbf{C}_x = \sum \bigotimes_{j=1}^{n-1} \mathbf{C}_{x_j} \otimes [\mathbf{K} + (\mathbf{I}-\mathbf{K})]$$

where the second summation is over all binary numbers $(x_1 x_2 \ldots x_{n-1})$. The result $\sum \mathbf{C}_x = \mathbf{I}$, implied by (6.10), can then be verified by induction.

Note also that all $\mathbf{C}_x$ matrices, apart from $\mathbf{C}_0$, are contrast matrices. This follows from the orthogonality property, as then $\mathbf{C}_x \mathbf{C}_0 = \mathbf{0}$ implies $\mathbf{C}_x \mathbf{1} = \mathbf{0}$ for all $x \neq 0$. Finally, since $\mathbf{C}_x$ is idempotent,

$$\operatorname{rank}(\mathbf{C}_x) = \operatorname{trace}(\mathbf{C}_x) = \prod_{j=1}^{n} \operatorname{trace}(\mathbf{C}_{x_j})$$

and the trace of $\mathbf{C}_{x_j}$ is 1 if $x_j = 0$, and $m_j - 1$ if $x_j = 1$. Hence

$$\operatorname{rank}(\mathbf{C}_x) = \prod_{j=1}^{n} (m_j - 1)^{x_j} \tag{6.12}$$

which gives the degrees of freedom associated with the generalized interaction $\mathbf{C}_x \boldsymbol{\tau}$.

Although (6.10) represents the treatment structure of an n-factor experiment, not all generalized interactions need be included in the fitted model. It is often the case in practice that only a subset of these interactions are considered for inclusion. In single replicate designs some effects, often the higher-order interactions, are usually assumed to be negligible and are used to provide an estimate of experimental error.

6.3 Randomized block designs

Suppose that the $v = m_1 m_2 \ldots m_n$ treatment combinations of an n-factor experiment having factor F_i at m_i levels $(i = 1, 2, \ldots, n)$ is set out in a randomized block design with $b = r$ blocks. The design consisting of a single block accommodating all treatment combinations is a special case with $r = 1$. To avoid duplication, results will now be given which apply when $r \geqslant 1$, although the use of multiple replication designs will be discussed in Chapter 7.

The analysis of a randomized block design is given in Section 3.2. The covariances between the estimators $\mathbf{C}_x\hat{\boldsymbol{\tau}}$ and $\mathbf{C}_y\hat{\boldsymbol{\tau}}$ of two different generalized interactions are

$$\operatorname{cov}(\mathbf{C}_x\hat{\boldsymbol{\tau}}, \mathbf{C}_y\hat{\boldsymbol{\tau}}) = \mathbf{C}_x\boldsymbol{\Omega}\mathbf{C}_y\sigma^2 = \mathbf{C}_x\mathbf{C}_y(\sigma^2/r) = \mathbf{0} \tag{6.13}$$

for all $x \neq y$, since from (1.42) $\boldsymbol{\Omega} = (1/r)\mathbf{I}$. Hence, the estimate of any contrast in the factorial effect $\mathbf{C}_x\boldsymbol{\tau}$ is orthogonal to the estimate of every contrast in the effect $\mathbf{C}_y\boldsymbol{\tau}$. Factorial effects are then said to be orthogonal to each other. This means that any factorial effect can be estimated independently of the other effects and without regard to which other effects have been included in the model. The adjusted treatment sum of squares can, in consequence, be partitioned orthogonally into sums of squares corresponding to each of the factorial effects.

From results in Section 3.2, the treatment contrast $\mathbf{C}_x\boldsymbol{\tau}$ is estimable with estimator

$$\mathbf{C}_x\hat{\boldsymbol{\tau}} = (1/r)\mathbf{C}_x\mathbf{T} \tag{6.14}$$

where $\mathbf{T}$ is the vector of treatment totals. Thus estimates are given by contrasts in the unadjusted treatment means. The variance–covariance matrix of this estimator is given by

$$V(\mathbf{C}_x\hat{\boldsymbol{\tau}}) = \mathbf{C}_x\boldsymbol{\Omega}\mathbf{C}_x\sigma^2 = \mathbf{C}_x(\sigma^2/r) \tag{6.15}$$

The sum of squares due to testing the hypothesis $H_0\!:\mathbf{C}_x\boldsymbol{\tau} = \mathbf{0}$ is, from (1.35),

$$S(H_0) = (\mathbf{C}_x\hat{\boldsymbol{\tau}})'(\mathbf{C}_x\boldsymbol{\Omega}\mathbf{C}_x)^-(\mathbf{C}_x\hat{\boldsymbol{\tau}})$$

i.e., using (6.14) and (6.15) and with $\mathbf{C}_x^- = \mathbf{I}$,

$$S(H_0) = (1/r)(\mathbf{C}_x\mathbf{T})'(\mathbf{C}_x\mathbf{T}) \tag{6.16}$$

This sum of squares is based on v degrees of freedom, where $v = \operatorname{rank}(\mathbf{C}_x)$ is given by (6.12).

The calculation of the sum of squares in (6.16) is particularly straightforward. To illustrate this, consider a factorial experiment with three factors F_1, F_2 and F_3 at m_1, m_2 and m_3 levels respectively. Let the treatment total corresponding to the treatment combination with factor F_j at level i_j be denoted by $T_{i_1i_2i_3}$ $(0 \leqslant i_j \leqslant m_j - 1; j = 1,2,3)$ and let

$$\bar{T}_{i_1i_2.} = \sum_{i_3} T_{i_1i_2i_3}/m_3, \qquad \bar{T}_{i_1..} = \sum_{i_2}\sum_{i_3} T_{i_1i_2i_3}/(m_2m_3)$$

and so on. For the main effect of factor F_1,

$$\mathbf{C}_{100}\mathbf{T} = [(\mathbf{I}-\mathbf{K})\otimes\mathbf{K}\otimes\mathbf{K}]\mathbf{T} = \bar{\mathbf{T}}_{F_1} - \bar{T}_{...}\mathbf{1}$$

where $\bar{\mathbf{T}}_{F_1}$ is a vector consisting of m_2m_3 replications of each of the m_1 elements $\bar{T}_{i_1..}$. Hence, the sum of squares due to the main effect of factor F_1 is

$$S(F_1) = (m_2m_3/r)\sum_{i_1=0}^{m_1-1}(\bar{T}_{i_1..} - \bar{T}_{...})^2$$

$$= \sum_{i_1=0}^{m_1-1} T_{i_1..}^2/(rm_2m_3) - \text{CF} \tag{6.17}$$

where $\text{CF} = T_{...}^2/(rm_1m_2m_3)$ is the correction factor for the overall mean. Hence, $S(F_1)$ is equal to the sum of squares of factor F_1 totals divided by the number of observations making up each of these totals, less the overall correction factor. It is the same type of formula as that obtained from fitting a set of parameters in a one-way analysis of variance model; see (1.29) or (1.32).

To obtain the F_1F_2 interaction sum of squares $S(F_1F_2)$ let

$$S(F_1 \text{ and } F_2) = \sum_{i_1}\sum_{i_2} T_{i_1i_2.}^2/(rm_3) - \text{CF}$$

Then, since

$$\mathbf{C}_{110} = (\mathbf{I} - \mathbf{C}_{00} - \mathbf{C}_{10} - \mathbf{C}_{01})\otimes\mathbf{K}$$
$$= (\mathbf{I}\otimes\mathbf{I}\otimes\mathbf{K}) - (\mathbf{K}\otimes\mathbf{K}\otimes\mathbf{K}) - \mathbf{C}_{100} - \mathbf{C}_{010}$$

it follows that

$$S(F_1F_2) = S(F_1 \text{ and } F_2) - S(F_1) - S(F_2)$$

The extension to higher factor interactions and to the general factorial experiment should be apparent.

6.4 Identifying components of interaction

For symmetrical p^n designs, where p is a prime or prime power, Bose (1947) and Kempthorne (1947) divided the complete set of treatment degrees of freedom into homogeneous orthogonal subsets; a subset is said to be homogeneous if all its contrasts belong to the same generalized interaction, and two subsets are orthogonal if all contrasts in one are orthogonal to those in the other. For example, it was shown in Section 6.1 that the F_1F_2 interaction in a 3^2 experiment could be divided into two orthogonal components, denoted by $F_1F_2^2$ and F_1F_2, each based on two degrees of freedom representing contrasts among treatment combinations which satisfied, respectively, the equations

$$a_1 + 2a_2 = 0, 1, 2 \qquad (\text{mod } 3)$$
$$a_1 + a_2 = 0, 1, 2 \qquad (\text{mod } 3)$$

These ideas were extended to include asymmetrical and other symmetrical designs by Bailey, Gilchrist and Patterson (1977).

First $H(z)$ is defined; it comprises a subset of degrees of freedom associated with the vector $z = z_1z_2\ldots z_n$ where z_i is one of the integers $0, 1, \ldots, m_i - 1$. A treatment combination is again represented by $a = a_1a_2\ldots a_n$, where a_i is a level of factor F_i $(i = 1, 2, \ldots, n)$. Let the integer $a(z)$ be given by

$$a(z) = \sum_{i=1}^{n} \delta_i a_i z_i \qquad (\text{mod } \gamma) \tag{6.18}$$

where γ is the lowest common multiple of the m_i and where $\delta_i = \gamma/m_i$. Then $H(z)$ is defined to represent the set of all contrasts between treatment combinations with different values of $a(z)$.

For example, for the 3^2 experiment $\gamma = 3$ and $\delta_i = 1$ $(i = 1, 2)$ so that $a(12) = a_1 + 2a_2$ and $a(11) = a_1 + a_2$, mod 3. Hence, $H(12)$ and $H(11)$ correspond to the $F_1F_2^2$ and F_1F_2 components respectively of the F_1F_2 interaction.

Consider now a $2 \times 3 \times 6$ experiment, i.e. with three factors F_1, F_2 and F_3 at 2, 3 and 6 levels respectively. Since $\gamma = 6$ and $\delta_1 = 3$, $\delta_2 = 2$ and $\delta_3 = 1$,

$$a(z) = 3a_1z_1 + 2a_2z_2 + a_3z_3$$

The values of $a(z)$ for $z = 114$ are given in Table 6.1.

There are six different $a(114)$ values so that $H(114)$ represents five degrees of freedom. $H(114)$ is not, however, homogeneous. The contrast between treatment combinations with even and odd values of $a(114)$ is the single F_1 main effect degree of freedom. There are also two degrees of freedom from the interaction F_2F_3; these are given by the contrasts between pairs of values 0 and 3, 1 and 4, 2 and 5. The remaining two degrees of freedom are from the three-factor interaction. Neither are the $H(z)$ orthogonal; for instance, $H(114)$ and $H(011)$ can be shown to share the same degree of freedom from

Table 6.1 *Values of a(114) for 2 × 3 × 6 experiment.*

Levels of F_1	Levels of F_2	Levels of F_3 0	1	2	3	4	5
0	0	0	4	2	0	4	2
0	1	2	0	4	2	0	4
0	2	4	2	0	4	2	0
1	0	3	1	5	3	1	5
1	1	5	3	1	5	3	1
1	2	1	5	3	1	5	3

F_2F_3 but otherwise are different, whereas $H(022)$ is a proper subset of $H(114)$.

However from $H(z)$, sets $H_*(z)$ can be derived that are homogeneous, orthogonal and exhaustive, in the sense that between them they account for all treatment degrees of freedom. In the $2 \times 3 \times 6$ experiment the F_1 main effect in $H(114)$ is identifiable as $H(100)$ and the two degrees of freedom from F_2F_3 are $H(022)$. The two degrees of freedom for the three-factor interaction can therefore be identified as those contrasts in $H(114)$ that are orthogonal to $H(100)$ and $H(022)$. These two degrees of freedom are called $H_*(114)$.

Bailey et al. (1977) give a simple method for obtaining a complete set of $H_*(z)$. The set of all vectors z form an abelian group G, where addition of two z vectors consists of addition of the individual z_i modulo m_i. Each vector z generates a cyclic subgroup $\langle z \rangle$ given by

$$\langle z \rangle : 0, z, 2z, \ldots, (s-1)z$$

where s is the smallest non-zero integer having $sz = 0$. When the integer r is coprime to s, $\langle rz \rangle = \langle z \rangle$; otherwise $\langle rz \rangle$ is a proper subgroup of $\langle z \rangle$.

$H_*(z)$ is now defined to be the set of all contrasts that are orthogonal to those in $H(s_1 z), \ldots, H(s_h z)$, where $s_1, s_2, \ldots, s_h$ are the proper prime divisors of s. The number of degrees of freedom in $H_*(z)$ is the Euler function $\phi(s)$, the number of integers between 1 and s which are coprime to s. Bailey et al. (1977) show that the set of $H_*(z)$ are orthogonal, homogeneous and exhaustive, with the contrasts in $H_*(z)$ belonging to the interaction defined by the non-zero elements of z.

Hence, continuing the $2 \times 3 \times 6$ example,

$$\langle 114 \rangle : 000 \quad 114 \quad 022 \quad 100 \quad 014 \quad 122$$

Now $s=6$, $\phi(s)=2$ and $s_1=2$, $s_2=3$, giving $\langle s_1 z\rangle$ and $\langle s_2 z\rangle$ as respectively

$$\langle 022\rangle:\quad 000 \quad 022 \quad 014$$
$$\langle 100\rangle:\quad 000 \quad 100$$

$H_*(114)$, therefore, represents two degrees of freedom from the $F_1F_2F_3$ interaction. It can be shown that the remaining 8 degrees of freedom of this interaction correspond to components $H_*(111)$, $H_*(112)$, $H_*(113)$ and $H_*(115)$, each with 2 degrees of freedom.

For symmetrical p^n experiments, where p is a prime, the $H_*(z)$ components are identical to those given by Bose (1947) and Kempthorne (1947). This follows since $s=p$ and $\phi(s)=p-1$, so that $H_*(z)=H(z)$ with $p-1$ degrees of freedom. If p is a prime power, i.e. $p=q^m$ where q is prime, then each treatment factor can be replaced by m pseudo-factors each at q levels. The sets of degrees of freedom described by Bose (1947) are then unions of the $H_*(z)$.

6.5 Construction of single replicate designs

In an n-factor experiment with factor F_i at m_i levels, a treatment combination is represented by $a=a_1a_2\ldots a_n$ where a_i is a level of F_i ($0\leqslant a_i\leqslant m_i-1$; $i=1,2,\ldots,n$). Addition of two treatment combinations a and b is defined as

$$a+b=a_1a_2\ldots a_n+b_1b_2\ldots b_n=c_1c_2\ldots c_n=c \qquad (6.19)$$

where $c_i=a_i+b_i$, mod $m_i (i=1,\ldots,n)$. For some non-negative integer u let $ua=a+a+\cdots+a$ (u times), i.e. ua is obtained by multiplying each a_i by u and reducing modulo m_i whenever necessary.

Now in the $2\times 3\times 6$ experiment discussed in the previous section, a design confounding the 5 degrees of freedom in $H(114)$ is obtained by placing the six treatment combinations with $a(114)=3a_1+2a_2+4a_3=0$, mod 6, in the initial block; those with $a(114)=1$, mod 6, in the next block; and so on. More generally, the treatment combinations in the initial block of a design will satisfy $a(z_i)=0$, mod γ, for a set of vectors $z_1, z_2, \ldots$ where $a(z)$ is defined in (6.18). Now if treatment combinations a and b both satisfy this equation then so will treatment combination u_1a+u_2b, for any non-negative integers u_1 and u_2. This result leads to the following general method of constructing single replicate designs.

Consider the cyclic subgroup of treatment combinations

$$G: 0, a, 2a, \ldots, (q-1)a \tag{6.20}$$

where q, the order of the group, is the smallest integer such that $qa = 0$. G is an abelian group under addition defined by (6.19) and has q distinct elements. Let the q treatment combinations in G comprise the initial block of a design with the remaining blocks given by the *cosets* of G, obtained by adding other treatment combinations to those in G. The resulting design is necessarily a single replicate design. The treatment combination a is the *generator* of G and, hence, of the design.

Example 6.1
Consider a 3^2 experiment with generator $a = 11$. Then G: 00 11 22 and the design is

(00 11 22)
(01 12 20)
(02 10 21)

The design was given earlier in Section 6.1.

More generally, let a_i be a generator of a subgroup G_i of order $q_i (i = 1, 2, \ldots, p)$; then

$$G: G_1 \oplus G_2 \oplus \cdots \oplus G_p \tag{6.21}$$

is the direct sum of these p subgroups with general element given by

$$u_1 a_1 + u_2 a_2 + \cdots + u_p a_p \tag{6.22}$$

$(0 \leqslant u_i < q_i; i = 1, 2, \ldots, p)$. Assuming that the generators a_i have been chosen in such a way that no treatment combination occurs more than once in G, then G is a subgroup of order $q = \prod_{i=1}^{n} q_i$. This can be achieved in a sequential manner by ensuring that a_2 is not in G_1, a_3 is not in $G_1 \oplus G_2$, a_4 is not in $G_1 \oplus G_2 \oplus G_3$, and so on. The q elements of G will then constitute the initial block of a single replicate design with the remaining blocks given by the cosets of G.

Example 6.2
Consider a 4^2 experiment with two generators $a_1 = 02$ and $a_2 = 11$.

Then

$$G_1{:}00 \quad 02$$
$$G_2{:}00 \quad 11 \quad 22 \quad 33$$

so that

$$G{:}00 \quad 11 \quad 22 \quad 33 \quad 02 \quad 13 \quad 20 \quad 31$$

Since $q_1 = 2$ and $q_2 = 4$, the single replicate design obtained using G has 2 blocks of 8 given by

(00 02 11 13 20 22 31 33)
(01 03 12 10 21 23 32 30)

Example 6.3
Consider the asymmetrical $3 \times 4 \times 6$ factorial experiment with the two generators $a_1 = 101$ and $a_2 = 023$. Then

$$G_1{:}000 \quad 101 \quad 202 \quad 003 \quad 104 \quad 205$$
$$G_2{:}000 \quad 023$$

giving

$$G{:}\,000 \quad 101 \quad 202 \quad 003 \quad 104 \quad 205$$
$$023 \quad 124 \quad 225 \quad 020 \quad 121 \quad 222$$

Now $q_1 = 6$ and $q_2 = 2$, so that the resulting single replicate design has 6 blocks of 12 given by

(000 003 020 023 101 104 121 124 202 205 222 225)
(001 004 021 024 102 105 122 125 203 200 223 220)
(002 005 022 025 103 100 123 120 204 201 224 221)
(010 013 030 033 111 114 131 134 212 215 232 235)
(011 014 031 034 112 115 132 135 213 210 233 230)
(012 015 032 035 113 110 133 130 214 211 234 231)

The choice of appropriate generators will depend on the confounding scheme of the resulting design. The $3 \times 4 \times 6$ design in Example 6.3, for instance, may be of little practical use since one degree of freedom from the main effect of factor F_2 is confounded with blocks; this degree of freedom being the comparison of the first three blocks, which has the second factor at levels 0 and 2 only, with the remaining blocks.

Finally, it can be noted that the single replicate designs obtained by the above method belong to the class of n-cyclic designs considered in Section 4.5; they are in fact disconnected partial n-cyclic sets.

6.6 Analysis of single replicate designs

Let the rows of the incidence matrix $\mathbf{N}$ of the single replicate design be permuted so that the first k rows correspond to the k treatment combinations in the first block, the next k rows to the k treatment combinations in the second block, and so on. Then $\mathbf{N} = \mathbf{I}_b \otimes \mathbf{1}_k$ and $\mathbf{NN}' = \mathbf{I}_b \otimes \mathbf{J}_k$, where $b = v/k$ is the number of blocks. Hence, the information matrix $\mathbf{A}$ of (1.12) is given by

$$\mathbf{A} = \mathbf{I} - (1/k)\mathbf{NN}' = \mathbf{I}_b \otimes (\mathbf{I}_k - \mathbf{K}_k) \tag{6.23}$$

Two important results follow directly from the fact that this information matrix is an idempotent matrix.

Firstly, the eigenvalues of $\mathbf{A}$ are either 0 or 1, which means that the canonical efficiency factors of a single replicate design are all equal to unity. In other words, if a treatment contrast is estimable than it is estimated with full efficiency.

Secondly, a generalized inverse of $\mathbf{A}$ is given by $\mathbf{\Omega} = \mathbf{I}$. Hence the analysis of a single replicate design follows along the lines of that of a randomized block design, with estimates of generalized interactions given by (6.14) and corresponding sum of squares by (6.16), where $r = 1$ and where the vector of unadjusted treatment totals $\mathbf{T}$ is now replaced by the vector of adjusted treatment totals $\mathbf{q}$; this vector being given by subtracting from each observation (since $r = 1$) its corresponding block mean. These estimates and sums of squares are based on the number of unconfounded degrees of freedom associated with the generalized interactions. If an interaction, as opposed to components of an interaction, is totally confounded then its estimate, sum of squares and degrees of freedom will be zero. If, on the other hand, an interaction is completely unconfounded the analysis proceeds exactly as for a randomized block design, since block mean adjustments will cancel each other out.

For confounded main effects and interactions it must be remembered that not all components can be estimated. For instance, consider a four-level factor which has the comparison of levels 0 and 1 with levels 2 and 3 confounded. Now this contrast cannot be estimated free of block effects, so that the main effect sum of squares will not include a contribution from this contrast. If this contrast is likely to be of importance then the design will clearly be unsatisfactory.

It now remains to be able to determine the confounding scheme of any given single replicate design constructed by the methods of Section 6.5.

6.7 Determining the confounding scheme

It has already been stated in Section 6.5 that if a design is required which confounds the subsets of degrees of freedom given by $H(z_1)$, $H(z_2),\ldots,$for some vectors $z_1, z_2, \ldots.$. then the initial block will contain those treatment combinations satisfying $a(z_i) = 0$, mod $\gamma(i = 1, 2, \ldots)$, where $a(z_i)$ is defined in (6.18). Given a single replicate design, however, the reverse process has to be carried out. The components of interaction confounded are identified by finding z vectors such that the treatment combinations in the initial block all provide integers defined in (6.18) equal to zero.

Example 6.4

Consider the following single replicate design for a 6^3 experiment in 6 blocks of 36 obtained from the two generators $a_1 = 012$ and $a_2 = 113$. A vector $z = z_1 z_2 z_3$ must now be found which satisfies both $a_1(z) = 0$ and $a_2(z) = 0$, i.e.

$$z_2 + 2z_3 = 0 \quad \text{and} \quad z_1 + z_2 + 3z_3 = 0 \qquad (\text{mod}\, 6)$$

Clearly $z_1 + z_3$ must be zero so that $z_1 = 1$ and $z_3 = 5$ gives $z_2 = 2$. Now

$$\langle 125 \rangle : 000 \quad 125 \quad 244 \quad 303 \quad 422 \quad 541$$

giving $s = 6$, $\phi(s) = 2$, and

$$\langle 244 \rangle : 000 \quad 244 \quad 422$$
$$\langle 303 \rangle : 000 \quad 303$$

Hence, the confounded components are $H_*(125)$ and $H_*(244)$ each with 2 degrees of freedom and $H_*(303)$ with 1 degree of freedom.

Example 6.5

Consider the design given in Example 6.3 of a $3 \times 4 \times 6$ experiment with the two generators $a_1 = 101$ and $a_2 = 023$. The design has 6 blocks so that 5 degrees of freedom will be confounded with blocks.

Now $z = z_1 z_2 z_3$ must satisfy

$$\begin{aligned} a_1(z) &= 4z_1 \qquad + 2z_3 = 0 \pmod{12} \\ a_2(z) &= \qquad 6z_2 + 6z_3 = 0 \pmod{12} \end{aligned}$$

The equations are satisfied by $z = 222$. Now

$$\langle 222 \rangle : 000 \quad 222 \quad 104 \quad 020 \quad 202 \quad 124$$

giving $s = 6, s_1 = 2, s_2 = 3, \phi(s) = 2$, and

$$\langle 104 \rangle : 000 \quad 104 \quad 202$$
$$\langle 020 \rangle : 000 \quad 020$$

Hence, the confounded components are $H_*(222)$ and $H_*(104)$ with 2 degrees of freedom and $H_*(020)$ with one degree of freedom. Thus, components of the main effect of F_2 and the F_1F_3 and $F_1F_2F_3$ interactions have been totally confounded in this design.

In both of the above examples, since there are a small number of blocks and a relatively large block size, the confounded components $H_*(z)$ can be readily found. Consider, however, a 6^3 design in 36 blocks of 6 obtained from the generator $a = 113$; i.e. with a large number of blocks but small block size. Finding the components $H_*(z)$ now becomes more tedious since there are many z vectors satisfying $z_1 + z_2 + 3z_3 = 0$, mod 6. For instance, it is satisfied by $z = 301$ so that it can consequently be shown that 3 degrees of freedom from the F_1F_3 interaction are confounded, which are identified by the components $H_*(301)$ and $H_*(303)$ with 2 and 1 degrees of freedom respectively. Although all components can be found in this way, it is sometimes easier to first determine the confounding scheme using the following alternative method.

Consider an n-factor single replicate design constructed according to the methods of Section 6.5. Since these designs are n-cyclic sets, it follows from (4.8) that the concurrence matrix $\mathbf{NN}'$ can be written as

$$\mathbf{NN}' = \sum_{a_1}\sum_{a_2}\cdots\sum_{a_n} \lambda_{a_1a_2\ldots a_n}(\mathbf{\Gamma}_{a_1} \otimes \mathbf{\Gamma}_{a_2} \otimes \cdots \otimes \mathbf{\Gamma}_{a_n}) \qquad (6.24)$$

where $\mathbf{\Gamma}_{a_i}$ is a basic circulant matrix as defined in Section 4.5 and where $\lambda_{a_1a_2\ldots a_n} = 1$ if treatment combination $a_1a_2\ldots a_n$ occurs in the initial block, and zero otherwise. Now an eigenvector of $\mathbf{NN}'$ is given by

$$\mathbf{h}_x = \mathbf{h}_{x_1} \otimes \mathbf{h}_{x_2} \otimes \cdots \otimes \mathbf{h}_{x_n}$$

where $\mathbf{h}_{x_i}$ is a vector of ones if $x_i = 0$, and is a contrast vector satisfying $\mathbf{h}'_{x_i}\mathbf{1} = 0$ if $x_i = 1$. Further, since $\mathbf{C}_{x_i}\mathbf{h}_{x_i} = \mathbf{h}_{x_i}$ for $x_i = 0$ or 1, where $\mathbf{C}_{x_i}$ is given by (6.9), then $\mathbf{h}_x$ is an eigenvector of $\mathbf{C}_x$ and is therefore in the vector space spanned by the columns of $\mathbf{C}_x$.

Hence there exists a set of contrast vectors which span the generalized interaction space defined by the columns of $\mathbf{C}_x$ and which are also eigenvectors of the information matrix $\mathbf{A}$. The following theorem can now be proved.

Theorem 6.1

The number of degrees of freedom of the generalized interaction $\mathbf{C}_x\boldsymbol{\tau}$ confounded with blocks is

$$Y_x = (1/k)\,\text{trace}(\mathbf{NN}'\mathbf{C}_x) \tag{6.25}$$

Proof Let the $v \times v$ matrix $\mathbf{C}_x$ have rank $t < v$ and let $\mathbf{H}_x$ be a $v \times t$ matrix of rank t whose columns are eigenvectors of $\mathbf{A}$, i.e. $\mathbf{AH}_x = \mathbf{H}_x\boldsymbol{\lambda}^\delta$ where $\boldsymbol{\lambda}^\delta$ is a diagonal matrix with elements $\lambda_1, \lambda_2, \ldots, \lambda_t$. The eigenvalue λ_i is equal to 0 or 1 since $\mathbf{A}$ is idempotent $(i = 1, 2, \ldots. t)$. Now since $\mathbf{H}'_x\boldsymbol{\tau}$ is estimable if and only if $\mathbf{H}_x = \mathbf{AH}_x$ (see Section 1.4) then the number of linearly independent estimable contrasts in $\mathbf{H}'_x\boldsymbol{\tau}$ is $\sum_{i=1}^{t}\lambda_i$. Let the columns of $\mathbf{H}_x$ be in the vector space spanned by the columns of $\mathbf{C}_x$ so that $\mathbf{H}_x = \mathbf{C}_x\mathbf{H}_x$. Also let $\mathbf{G}_x$ be a $v \times (v - t)$ matrix of rank $v - t$ such that $\mathbf{C}_x\mathbf{G}_x = \mathbf{0}$. It then follows that

$$\mathbf{AC}_x(\mathbf{H}_x \vdots \mathbf{G}_x) = (\mathbf{H}_x \vdots \mathbf{G}_x)\begin{pmatrix} \boldsymbol{\lambda} \\ \mathbf{0} \end{pmatrix}^\delta$$

so that

$$\sum_{i=1}^{t} \lambda_i = \text{trace}(\mathbf{AC}_x)$$

This represents the number of degrees of freedom of the generalized interaction $\mathbf{C}_x\boldsymbol{\tau}$ that are estimable. Hence, the number of degrees of freedom confounded is

$$t - \text{trace}(\mathbf{AC}_x) = \text{trace}(\mathbf{C}_x) - \text{trace}(\mathbf{AC}_x) = Y_x$$

which completes the proof.

For the designs constructed by the methods of Section 6.5, (6.25) simplifies to give

$$Y_x = (1/k)\sum\left(\prod_j w_{a_j}\right) \tag{6.26}$$

where the summation is over all treatment combinations in the initial block, the product is over all j for which $x_j = 1$, and where

$$w_{a_j} = \begin{cases} m_j - 1 & \text{if } a_j = 0 \\ -1 & \text{if } a_j \neq 0 \end{cases}$$

With $\mathbf{NN}'$ given by (6.24) it follows that

$$\text{trace}(\mathbf{NN}'\mathbf{C}_x) = \sum \prod_{j=1}^{n} \text{trace}(\mathbf{\Gamma}_{a_j}\mathbf{C}_{x_j})$$

If $x_j = 0$ then $\mathbf{\Gamma}_{a_j}\mathbf{C}_{x_j} = \mathbf{K}$ so that its trace is 1. If $x_j = 1$ then $\mathbf{\Gamma}_{a_j}\mathbf{C}_{x_j} = \mathbf{\Gamma}_{a_j} - \mathbf{K}$ and

$$\text{trace}(\mathbf{\Gamma}_{a_j} - \mathbf{K}) = \begin{cases} m_j - 1 & \text{if } a_j = 0 \\ -1 & \text{if } a_j \neq 0 \end{cases}$$

The result given in (6.26) is thus established.

A number of examples will now be given to illustrate how Y_x in (6.26) is calculated. Results can be conveniently set out in tabular form with the treatment combinations in the initial block in the first column and the w_{a_j} and their products in the other columns.

Example 6.6
For the 3^2 experiment of Example 6.1 with generator $a = 11$, the calculations are as set out in Table 6.2.

Hence

$$Y_{10} = (1/3)\sum w_{a_1} = 0, \qquad Y_{01} = (1/3)\sum w_{a_2} = 0,$$
$$Y_{11} = (1/3)\sum w_{a_1} w_{a_2} = 2$$

Table 6.2 *Calculations for confounding in 3^2 experiment.*

Initial block $a_1 a_2$	w_{a_1}	w_{a_2}	$w_{a_1} w_{a_2}$
00	2	2	4
11	−1	−1	1
22	−1	−1	1
Total	0	0	6

Table 6.3 *Calculations for confounding in 3 × 4 × 6 experiment.*

Initial block	w_{a_1}	w_{a_2}	w_{a_3}	$w_{a_1}w_{a_2}$	$w_{a_1}w_{a_3}$	$w_{a_2}w_{a_3}$	$w_{a_1}w_{a_2}w_{a_3}$
000	2	3	5	6	10	15	30
003	2	3	−1	6	−2	−3	−6
020	2	−1	5	−2	10	−5	−10
023	2	−1	−1	−2	−2	1	2
101	−1	3	−1	−3	1	−3	3
104	−1	3	−1	−3	1	−3	3
121	−1	−1	−1	1	1	1	−1
124	−1	−1	−1	1	1	1	−1
202	−1	3	−1	−3	1	−3	3
205	−1	3	−1	−3	1	−3	3
222	−1	−1	−1	1	1	1	−1
225	−1	−1	−1	1	1	1	−1
Total	0	12	0	0	24	0	24
Y_x	0	1	0	0	2	0	2

confirming that 2 degrees of freedom from the F_1F_2 interaction are confounded with blocks

Example 6.7
For the single replicate design of Example 6.3 for a 3 × 4 × 6 experiment with generators $a_1 = 101$ and $a_2 = 023$ the calculations are given in Table 6.3. These confirm that 1 degree of freedom from the F_2 main effect and 2 degrees of freedom from the F_1F_3 and $F_1F_2F_3$ interactions are confounded with blocks.

Example 6.8
Returning to the 6^3 experiment in 36 blocks of 6 obtained from the generator $a = 113$, the results are given in Table 6.4.

The use of (6.26) gives a simple way of obtaining the number of degrees of freedom confounded with blocks for any generalized interaction. It does not identify these degrees of freedom with the components $H_*(z)$ of the interaction but, if it is necessary to specify these components, this can be done subsequently. This task is in

Table 6.4 *Calculations for confounding in 6^3 experiment.*

Initial block	w_{a_1}	w_{a_2}	w_{a_3}	$w_{a_1}w_{a_2}$	$w_{a_1}w_{a_3}$	$w_{a_2}w_{a_3}$	$w_{a_1}w_{a_2}w_{a_3}$
000	5	5	5	25	25	25	125
113	−1	−1	−1	1	1	1	−1
220	−1	−1	5	1	−5	−5	5
333	−1	−1	−1	1	1	1	−1
440	−1	−1	5	1	−5	−5	5
553	−1	−1	−1	1	1	1	−1
Total	0	0	12	30	18	18	132
Y_x	0	0	2	5	3	3	22

any case simplified when the general confounding scheme is known, especially when there are a large number of blocks.

When the number of levels of a factor is not a prime number then the use of *pseudo-factors* can often be used to enlarge the class of available designs. More specifically, Voss and Dean (1987) have shown that when the number of levels of a factor has a prime-powered divisor then different designs may sometimes be obtained by the use of pseudo-factors. The number of levels associated with each pseudo-factor will be a prime number. Suppose, for instance, that a factor F_i at 4 levels is represented by two pseudo-factors F_{i1} and F_{i2} each at two levels. The correspondence between levels of F_i and F_{i1} and F_{i2} will be as shown in Table 6.5. The 3 degrees of freedom of the main effect of F_i correspond to the main effects F_{i1} and F_{i2} and the interaction $F_{i1}F_{i2}$, each with a single degree of freedom. The degrees of freedom of the interaction of factor F_i with factor $F_j (j \neq i)$ will be identified with those of the interactions $F_{i1}F_j$, $F_{i2}F_j$ and $F_{i1}F_{i2}F_j$. Other interactions will be given in a similar way.

Table 6.5 *Correspondence between levels of factor and pseudo-factors.*

Factor	*Level*			
F_i	0	1	2	3
F_{i1}	0	0	1	1
F_{i2}	0	1	0	1

Using the pseudo-factors, designs can now be constructed by the methods of Section 6.5. The confounding scheme of the pseudo-factor design can be determined by the above methods. Although care has to be taken in identifying the interactions confounded in the pseudo-factor design with those in the corresponding design for the original factors, the procedure presents no real difficulties. An example will illustrate the technique.

Example 6.9
Consider a factorial experiment with four factors F_1, F_2, F_3 and F_4 at 2, 3, 4 and 6 levels respectively. Suppose a single replicate design in 12 blocks of 12 is required. Let factor F_3 be represented by two pseudo-factors F_{31} and F_{32} each at two levels. For the $2 \times 3 \times (2 \times 2) \times 6$ experiment, the two generators $a_1 = 11101$ and $a_2 = 00013$ produce a design of the required size, which confounds 1 degree of freedom from F_1F_{31}, $F_{31}F_{32}F_4$ and $F_1F_{32}F_4$, and 2 degrees of freedom from F_2F_4, $F_2F_{31}F_{32}F_4$, $F_1F_2F_{31}F_4$ and $F_1F_2F_{32}F_4$, i.e. 1 degree of freedom from F_1F_3, F_3F_4 and $F_1F_3F_4$, 2 degrees of freedom from F_2F_4 and $F_2F_3F_4$ and 4 degrees of freedom from $F_1F_2F_3F_4$. If pseudo-factors are not used then it is not possible to construct a design with more than 2 degrees of freedom from the four-factor interaction $F_1F_2F_3F_4$ confounded with blocks. Giovagnoli (1977), in considering this example, also lets factor F_4 at 6 levels be represented by two pseudo-factors F_{41} and F_{42} at three and two levels respectively. She produces a design with the same confounding scheme as that above; its generators being $a_1 = 111001$ and $a_2 = 000110$. The result given by Voss and Dean (1987) shows, however, that it is unnecessary to use pseudo-factors for F_4 as designs with different confounding properties cannot be produced by doing so.

Finally, it should be noted that, given the generators, it is a relatively straightforward task to write a computer program which will determine the confounding scheme of a single replicate design using the above methods. It is particularly simple to write a program to calculate Y_x given in (6.26). First, the initial block of the design is obtained from (6.20) and (6.21) by constructing the cyclic subgroup G_i for each generator $a_i (i = 1, 2, \ldots, p)$ and then taking the direct sum of these subgroups. Next, the quantities w_{a_j} can be calculated for each factor in turn for every treatment combination in the initial block. Finally, the number of degrees of freedom confounded with blocks for each generalized interaction can be calculated from Y_x.

Determining which components of the generalized interactions are confounded can also be programmed without too much difficulty. The main steps in the program are as follows:

1. Obtain the set z of all z-vectors satisfying $a_i(z) = 0$ for $i = 1, 2, \ldots, p$.
2. Find the order of each cyclic subgroup generated from every element in z.
3. From z take a z-vector, z_1 say, corresponding to a group of lowest order. Then $H_*(z_1)$ is confounded, with degrees of freedom given by the number of non-zero elements in the group. Now delete all z-vectors belonging to this group from z.
4. From the remaining elements in z, take a z-vector corresponding to a group of lowest order. This gives another confounded component with degrees of freedom given by the number of non-zero z-vectors in this group which are still in z. All elements belonging to this group are then deleted from z. The procedure is repeated until all components have been found.

For instance, Example 6.4 would give

z :	125	244	303	422	541
Group order :	6	3	2	3	6

and the algorithm would produce the confounding scheme given in the example.

The availability of such a computer algorithm together with a careful choice of generators would enable a design of the required size and with the required properties, assuming such a design exists, to be quickly found.

6.8 Choice of generators

The size of the experiment and the required confounding scheme determine the choice of generators for the single replicate design. Appropriate generators can usually be found quickly by trial-and-error methods, as the confounding pattern of any given set of generators is easily obtained. Of course, the constraints imposed by the size and required confounding scheme may be such that no suitable design can be found.

Simple rules for choosing generators can be derived from (6.26) to ensure that all main effects are unconfounded, which is usually an essential requirement. For instance, suppose a design is to be constructed using a single generator $a = a_1 a_2 \ldots a_n$ where the ith factor F_i is at m_i levels $(i = 1, 2, \ldots, n)$. Then the main effect F_i is unconfounded if and only if a_i is relatively prime to m_i, i.e. $(m_i, a_i) = 1$. In this case,

$$g_i{:}0 \quad a_i \quad 2a_i \quad \ldots \quad (m_i - 1)a_i \tag{6.27}$$

constitutes a group closed to addition modulo m_i. Each zero must, therefore, be accompanied in the initial block by $m_i - 1$ non-zero elements. Hence, the sum $\sum w_{a_i}$ in (6.26) is zero.

With more than one generator it is only necessary for m_i to be relatively prime to the ith element in one of the generators for the F_i main effect to be unconfounded. Again this follows from (6.27) and (6.26). For instance, with two generators $a = a_1 a_2 \ldots a_n$ and $b = b_1 b_2 \ldots b_n$ the F_i main effect will be unconfounded if and only if $(m_i, a_i, b_i) = 1$. Consider, as an example, the $3 \times 4 \times 6$ experiment with generators $a_1 = 101$ and $a_2 = 023$ given in Example 6.3. The F_2 main effect is confounded since $(4, 0, 2) = 2$, but the F_1 and F_3 main effects are not since $(3, 1, 0) = 1$ and $(6, 1, 3) = 1$.

For a two-factor interaction to be unconfounded it is necessary that, for the treatment combinations in the initial block, each distinct level of one factor should be accompanied by every level of the other factor. For example, the design for a 4^2 experiment with generators $a_1 = 21$ and $a_2 = 22$ has initial block

$$00 \quad 21 \quad 02 \quad 23 \quad 22 \quad 03 \quad 20 \quad 01$$

The four treatment combinations with factor F_1 at level 0 have factor F_2 at each of its four levels; similarly for F_1 at level 2. Hence, the $F_1 F_2$ interaction will be unconfounded. Again, for the $2 \times 3 \times 4$ design with generator $a = 111$ it can be readily verified from the 12 treatment combinations in the initial block that $F_1 F_2$ and $F_2 F_3$ interactions are unconfounded.

One particular case when the $F_i F_j$ interaction, say, is necessarily unconfounded is where at least one of the generators has its ith element zero and jth element non-zero and relatively prime to m_j, or vice-versa. For instance, in the $3 \times 4 \times 6$ design of Example 6.3, inspection of the two generators immediately shows that $F_1 F_2$ and $F_2 F_3$ are unconfounded. For a further example, a 4^3 experiment in 16 blocks of 4 can be obtained using two generators both of order 4.

Using the above results, it can be readily established that use of generators $a_1 = 101$ and $a_2 = 011$ will ensure that all main effects and all two-factor interactions are unconfounded.

For three-factor interactions each distinct pair of levels of two of the factors must be accompanied by every level of the third factor. The extension to any interaction is obvious, although direct application of the methods in Section 6.7 would probably be more appropriate for interactions involving more than two factors.

6.9 2^n experiments

2^n factorial experiments involving n factors each at two levels are widely used in practice and have been extensively studied. A rigorous theory for constructing 2^n designs corresponding to particular confounding schemes is available and is given in most books on the design of experiments; see, for instance, John and Quenouille (1977, Chapter 5). This theory is a special case of that presented here for single replicate designs.

For a 2^n experiment a treatment combination $a = a_1 a_2 \ldots a_n$ will have $a_i = 0$ or 1 for all i, corresponding to, say, the absence or low level and presence or high level respectively of the factor. Each generator $a (a \neq 0)$ will, therefore, give rise to a group G consisting of just two elements 0 and a. The initial or *principal* block of a 2^n design consists of 0 and the elements of the direct sum of the generators. For example, suppose there are three distinct generators a, b and c, then the intial block will be

$$0,\ a,\ b,\ a+b,\ c,\ a+c,\ b+c,\ a+b+c$$

To ensure that no treatment combination occurs more than once in the initial block, $c \neq a + b$.

The conventional notation for a treatment combination in a 2^n experiment is to represent the high level of, say, factor A by its lower case letter a and the low level by the absence of that letter, with treatment combination $00\ldots0$ denoted by (1). For example, with four factors A, B, C, D,

$$1010 = ac, \qquad 0111 = bcd, \qquad 0010 = c, \qquad 0000 = (1)$$

The addition of two treatment combinations $a + b$ corresponds to multiplication in the conventional notation with the (modulo 2) rule that $a^2 = b^2 = c^2 = d^2 = 1$. For instance,

$$1010 + 0111 = 1101 \qquad \text{or} \qquad ac \times bcd = abc^2d = abd.$$

From the results of Section 6.4, the integer $a(z) = \sum a_i z_i \pmod 2$ can take one of two values, namely 0 or 1. Hence, the component $H(z)$ comprises one degree of freedom and corresponds to the generalized interaction $\mathbf{C}_x\boldsymbol{\tau}$ where $x_i = z_i$ for all i. Hence, $\mathbf{C}_x\boldsymbol{\tau}$ is confounded with blocks if and only if for all treatment combinations in the principal block

$$\sum_{i=1}^{n} a_i x_i = 0 \qquad \pmod 2 \tag{6.28}$$

In fact, it is only necessary for (6.28) to hold for the generators of the principal block, since if $\sum a_i x_i = 0$ and $\sum b_i x_i = 0$ then $\sum (a_i + b_i)x_i = 0$, all modulo 2.

The condition given by (6.28) is the 'even–odd' rule usually used for choosing the principal block. Consider, for instance, a 2^4 experiment with factors A, B, C, D and suppose ABC is to be confounded. Then (6.28) becomes $a_1 + a_2 + a_3 = 0$, mod 2. Hence, an even or zero number of the a_i $(i = 1, 2, 3)$ must be equal to 1. Using the conventional notation, each generator must have an even (or zero) number of letters in common with ABC. Possible generators are, for example, 0001, 1101 or 1010, i.e. d, abd and ac.

Similar results follow if more than one effect is to be confounded. It follows from (6.28) that if $\mathbf{C}_x\boldsymbol{\tau}$ and $\mathbf{C}_y\boldsymbol{\tau}$ $(x \neq y \neq 0)$ are confounded with blocks then so will $\mathbf{C}_{x+y}\boldsymbol{\tau}$. For example, suppose ABC and ACD are confounded in the 2^4 experiment; then $x = 1110$ and $y = 1011$ so that $x + y = 0101$, showing that BD is also confounded. Again, addition of effects corresponds to the multiplication of effects using the conventional notation with the rule that $A^2 = B^2 = C^2 = \cdots = 1$. Hence, $ABC \times ACD = A^2BC^2D = BD$. The treatment combinations in the principal block will have an even or zero number of letters in common with ABC and ACD (and also BD). The set of generalized interactions confounded together with the identity I are called the *defining contrasts* of the 2^n design and form a group closed to either addition or multiplication (mod 2) depending on whether the notation of this chapter or the conventional 2^n notation is used.

6.10 Fractional replication

The number of treatment combinations in a factorial experiment increases rapidly with the number of factors tested. Even with a single replicate design, the size of the experiment will frequently be too large to carry out in practice. If certain factorial effects, usually the high-order interactions, can be assumed to be negligible then it may be possible to use designs that require only a fraction of the total number of treatment combinations and which still permit the important effects – main effects and two-factor interactions, say – to be estimated. Such designs are called *fractional* factorial designs.

For example, consider a 3×2^5 experiment involving 96 treatment combinations. Suppose only main effects and two-factor interactions are of interest, with all other effects assumed to be negligible. Testing all 96 treatment combinations now becomes unnecessary since by taking a suitable one-half replicate design, involving 48 treatment combinations, it will be possible to estimate all main effects and two-factor interactions and still provide 20 degrees of freedom for error.

The fractional factorial designs considered here will consist of the treatment combinations in the initial block of the single replicate designs constructed by the methods of Section 6.5.

Consider the single replicate design for three factors F_1, F_2 and F_3 at 2, 3 and 4 levels respectively in 2 blocks of 12 obtained from the generator $a = 111$. The single degree of freedom confounded with blocks is from the F_1F_3 interaction, being in fact the component $H_*(102)$. A $\frac{1}{2}$-replicate design is obtained by taking the treatment combinations comprising the initial block of this single replicate design. The fractional factorial design consists, therefore, of the 12 treatment combinations

000	002	010	012	020	022
101	103	111	113	121	123

Since $H_*(102)$ is estimated in the single replicate design by the difference between the two block totals, then this component is completely lost in the fractional design. Now consider the single degree of freedom for the F_1 main effect, namely the $H_*(100)$ component. It represents the comparison between the first six treatment combinations in the design, having factor F_1 at level 0, with the remaining six treatment combinations having F_1 at level 1. However, this particular comparison also represents one degree of freedom from the F_3 main effect, the $H_*(002)$ component. It can be seen that the first six treatment combinations have factor F_3 at levels

0 and 2 while the remaining six have F_3 at levels 1 and 3. That is, $H_*(100)$ and $H_*(002)$ are estimated by the same contrast. These two components are then said to be *aliased*. It can be shown that the remaining two degrees of freedom of the F_3 main effect, $H_*(001)$, are aliased with two degrees of freedom from the F_1F_3 interaction, the $H_*(101)$ component. Thus, $H_*(001)$ and $H_*(101)$ are aliased. The aliases of the other components can also be found and, hence, the full aliasing system of the design determined.

An effect is estimable only if its aliases can be assumed to be negligible. To assess the usefulness of any fractional design it is, therefore, important that the aliasing system can be readily obtained.

Assume that in the single replicate design $H(z)$ is confounded with blocks. For each treatment combination a in the initial block the integer $a(z)$ defined by (6.18) will, therefore, equal zero. $H(z)$ is a defining contrast of the design. For $x \neq y \neq z$, $H_*(x)$ and $H_*(y)$ will be aliased if and only if they are estimated by the same contrasts, i.e. if and only if, for each treatment combination a in the initial block, $a(x) = a(y)$. In fact, it is only necessary that $a(x) = a(y)$ for the generating treatment combinations since if the equality holds for treatment combinations a and b it must also hold for $a + b$. For the $2 \times 3 \times 4$ example above,

$$a(z) = 6a_1z_1 + 4a_2z_2 + 3a_3z_3 \qquad (\text{mod } 12)$$

so with defining contrast $H(102)$,

$$a(102) = 6a_1 + 6a_3 \qquad (\text{mod } 12)$$

All 12 treatment combinations given above have $a(102) = 0$, mod 12. Also, since $a(100) = a(002)$ for the generator $a = 111$, $H_*(100)$ is aliased with $H_*(002)$ and, as $a(001) = a(103)$, $H_*(001)$ is aliased with $H_*(103) = H_*(101)$. The remaining aliases can be established in this way.

Consider the components $H_*(x)$ and $H_*(y)$ and suppose $y = x + t$ for some vector t, where addition of two vectors is defined in Section 6.4. Then, from (6.18)

$$a(y) = \sum_{i=1}^{n} \delta_i a_i(x_i + t_i) = a(x) + a(t) \qquad (\text{mod } \gamma) \qquad (6.29)$$

It follows that $a(y) = a(x)$ if and only if $a(t) = 0$, i.e. if and only if t is some multiple of the vector z. The aliases of $H_*(x)$ can, therefore, be determined from the subgroup given by the direct sum of the cyclic subgroups $\langle x \rangle$ and $\langle z \rangle$. Some examples will illustrate the procedure.

Example 6.10

Consider a 3^4 experiment in which $H(1111)$ is the defining contrast of a single replicate design in 3 blocks of 27. The corresponding factorial design will be a $\frac{1}{3}$-replicate with all treatment combinations satisfying $a_1 + a_2 + a_3 + a_4 = 0$, mod 3. The 27 treatment combinations can be obtained from the three generators 1110, 0111 and 1011. Now since

$$\begin{array}{lll} \langle 1000 \rangle + \langle 1111 \rangle : 0000 & 1111 & 2222 \\ 1000 & 2111 & 0222 \\ 2000 & 0111 & 1222 \end{array}$$

it follows that $H_*(1000)$ is aliased with $H_*(2111) = H_*(1222)$ and with $H_*(0222) = H_*(0111)$. The notation

$$H_*(1000) \equiv H_*(1222) \equiv H_*(0111)$$

will be used to show that they are aliased. Alternatively, using the notation of Section 6.1 the F_1 main effect is aliased with components $F_1F_2^2F_3^2F_4^2$ and $F_2F_3F_4$. It can be shown that all main effects are aliased with three- and four-factor interactions so that main effects are estimable if these interactions can be assumed to be negligible. However, some components of two-factor interactions are aliased with components of other two-factor interactions. For instance,

$$H_*(1100) \equiv H_*(0011) \equiv H_*(1122)$$

Other two-factor components are aliased with interactions involving three or more factors, for example

$$H_*(1200) \equiv H_*(0122) \equiv H_*(1022)$$

This $\frac{1}{3}$-replicate will provide a useful design if three- and four-factor interactions and some two-factor interactions can be assumed to be negligible. Alternatively (and this applies to all fractional factorials), the design could be used as an initial screening experiment in which the factorial effects likely to be of importance are identified, with another smaller experiment subsequently carried out to separate out any important aliased effects.

If the number of treatment combinations in a fractional factorial experiment is large then it may be necessary to set them out in a number of blocks. For instance, in the 3^4 factorial experiment of Example 6.10 the 27 treatment combinations could be set out in 3 blocks of 9 as follows. Suppose that the two degrees of freedom of the $H_*(1022)$ component of the $F_1F_3F_4$ interaction are confounded with blocks. Of course, since $H_*(1022)$ is aliased with $H_*(1200)$ and $H_*(0122)$ both of these components will also be confounded. The three blocks are obtained by arranging the 27 treatment combinations in the $\frac{1}{3}$-replicate to satisfy the equations $a(1022) = 0, 1, 2$ (mod 3). The generators of the initial block can be taken as 1110 and 1101, both of which satisfy $a(1111) = 0$ and $a(1022) = 0$, mod 3. The full design is:

(0000 1110 2220 1101 2211 0021 2202 0012 1122)
(0111 1221 2001 1212 2022 0102 2010 0120 1200)
(0222 1002 2112 1020 2100 0210 2121 0201 1011)

Setting out a fractional factorial design in blocks, therefore, presents no additional problems apart from the extra care needed in choosing the components to be confounded, since any components aliased with these chosen components will also be confounded.

Example 6.11

Now consider the $3 \times 4 \times 6$ factorial design in 6 blocks of 12 with generators $a_1 = 101$ and $a_2 = 023$ given in Example 6.3. The 5 degrees of freedom confounded with blocks was shown in Example 6.5 to be the subset $H(222)$, which was identified with components $H_*(222)$ and $H_*(104)$ with 2 degrees of freedom each and $H_*(020)$ with 1 degree of freedom. A $\frac{1}{6}$-replicate, with defining contrast $H(222)$, is given by the 12 treatment combinations in the initial block of this single replicate design, namely

000 003 020 023 101 104 121 124 202 205 222 225

Now the aliases of the F_1 main effect can be determined from the direct sum of the cyclic subgroups $\langle 100 \rangle$ and $\langle 222 \rangle$, which is

$\langle 100 \rangle + \langle 222 \rangle$: 000 222 104 020 202 124
100 022 204 120 002 224
200 122 004 220 102 024

Hence

$$H_*(100) \equiv H_*(002) \equiv H_*(022) \equiv H_*(120) \equiv H_*(102) \equiv H_*(122)$$

In a similar way it can be established that the $H_*(010)$ component of the F_2 main effect is aliased with the $H_*(114)$ component of the $F_1F_2F_3$ interaction. However, although the component $H_*(010)$ represents 2 degrees of freedom from the F_2 main effect, it is clear that only 1 degree of freedom is estimable in this fractional replicate; each treatment combination has the second factor at either level 0 or level 2. This degree of freedom is thus aliased with one of the four degrees of freedom of the $H_*(114)$ component. In general, the number of degrees of freedom of the $H_*(z)$ component which will be estimable in the fraction is determined by the number of distinct values given by the function $a(z)$ for the treatment combinations in the fraction. For this example,

$$a(z) = 4a_1z_1 + 3a_2z_2 + 2a_3z_3 \qquad (\text{mod } 12)$$

It can be verified that $a(010)$ takes values 0 or 6 for the above 12 treatment combinations, showing that only one degree of freedom of this main effect component is estimable.

Aliases of other effects can be similarly established using the direct sum of cyclic subgroups approach and the number of estimable degrees of freedom determined from the function $a(z)$. To set out the 12 treatment combinations in 2 blocks of 6 the single degree of freedom from $H_*(010) \equiv H_*(114)$ could be chosen to be confounded with blocks. As an alternative, the single degree of freedom from $H_*(013) \equiv H_*(111)$ could be used. This would give the design

(000 023 104 121 202 225)
(003 020 101 124 205 222)

Note that $a(013) = 0$ for all treatment combinations in the initial or principal block and $a(013) = 6$ for those in the other block, where all quantities are reduced modulo 12 where necessary.

Of course, these $\frac{1}{6}$-replicate designs in either a single block or in 2 blocks of 6 have little practical appeal since main effect components are either used in the defining contrasts, or are aliased with each other or are not estimable.

In a 2^n experiment, the above results show that if $H(z_i)$ is a defining contrast of the fractional factorial design then $H(x)$ is aliased with

$H(x + z_i)$ for all i. Using the conventional notation, aliases are readily established from the defining contrasts by use of the multiplication rule of Section 6.9. For example, in a 2^6 experiment with factors A, B, C, D, E and F, a $\frac{1}{4}$-replicate can be obtained using the defining contrasts

$$I = ABCD = ABEF = CDEF$$

so that

$$A \equiv BCD \equiv BEF \equiv ACDEF$$
$$AB \equiv CD \equiv EF \equiv ABCDEF$$

and so on, using on multiplication $A^2 = B^2 = \cdots = F^2 = 1$. If all interactions involving three or more factors are assumed negligible, then all main effects are estimable although some two-factor interactions will still be aliased with other two-factor interactions.

In Section 6.7 the outline of a computer program to find the confounding scheme of a single replicate design given its generators was described. Again, using the methods of this section, it is not difficult to add a subroutine to this program which would determine the aliasing scheme of the fractional design resulting from using only the initial block of the confounded design.

The analysis of fractional factorial designs follows along similar lines to that given in Section 6.6 for single replicate designs. Any component of a generalized interaction confounded in the single replicate design will now be completely lost in the fraction. The estimators and sums of squares of other components are obtained as in Section 6.6, where these quantities will necessarily be equal for any two components that are aliased.

6.11 Other designs

A variety of methods have been given for the construction of single replicate and fractional designs for factorial experiments. The earliest method was given by Bose and Kishen (1940) and by Bose (1947) and was applicable to symmetric p^n experiments, where p is a prime number. The method was later generalized by White and Hultquist (1965) to include experiments of the form $p^m q^n$, where p and q are distinct primes. Other generalizations were given by Raktoe (1969), Worthley and Banerjee (1974) and Sihota and Banerjee (1981). Cotter (1974) presented a method for constructing s^n designs, where s is not restricted to being a prime. All of these methods have, however, been shown by Voss and Dean (1987) to give designs which are identical to those produced by the method of Section 6.5.

Patterson (1965, 1976) describes an algorithm to construct designs and identify confounding patterns for factorial experiments, which is incorporated in a computer program called DSIGN. The algorithm can be used to obtain single replicate block designs in which every component $H_*(z)$ is either totally confounded, partially confounded or unconfounded. Fractional factorial block designs and designs using more complex blocking structures, such as row–column designs, can also be constructed. Bailey (1977) restricted attention from the very broad class of designs generated by the DSIGN algorithm to the subclass in which every component $H_*(z)$ is either completely confounded or is unconfounded. The method of construction given in Section 6.5 follows that given by Bailey (1977), although again the results of Bailey are more general than those presented here.

Finally, John and Dean (1975) and Dean and John (1975) have given a method of obtaining single replicate designs using the n-cyclic method of construction. As has already been stated, the designs produced by the method of Section 6.5 are in fact disconnected partial n-cyclic sets, and are thus identical to those given by Dean and John.

Single replicate and fractional designs having factors at two and three levels have been extensively tabulated; see, for instance, the three volumes produced by the National Bureau of Standards (1957, 1959, 1961) and recently reproduced in McLean and Anderson (1984); also Greenfield (1976, 1978), Box, Hunter and Hunter (1978), Fries and Hunter (1980) and Franklin (1984). Other designs have been tabulated by Dean and John (1975) and Lewis (1982).

Factorial experiments can also be set out in single and fractional replicate row–column designs; the way this can be done is considered in Section 7.8.

CHAPTER 7

Factorial experiments: multiple replication

7.1 Introduction

More precise estimates of treatment contrasts can be obtained if treatment combinations are replicated more than once in the factorial design. If the number of treatment combinations is not too large or if a large amount of experimental material is available, then multi-replicate factorial designs can often be employed. A further advantage is that an independent estimate of experimental error will be available from a comparison among replicate treatments. Such designs will be discussed in this chapter.

One method of constructing multi-replicate factorial designs is to take replications of the single replicate designs considered in the previous chapter. For instance, suppose a design is required for a 3^2 experiment in blocks of 3 with each treatment combination replicated twice. Two replicates of the design confounding component $F_1F_2^2$ of the F_1F_2 interaction could be used. No information will be available within blocks on this component but all other effects will be estimable free of block effects. An alternative would be to use single replicate designs which confound different components of the factorial effects; in this example, for instance, the $F_1F_2^2$ component could be confounded in one replicate and the F_1F_2 component in the second replicate. Now full information is still available on both main effects, and both components of the interaction can be estimated within blocks from half the observations. This interaction is said to be *partially confounded* with blocks. Such a design is frequently to be preferred since some information is available within blocks on all treatment contrasts.

The 3^2 design with two replicates confounding respectively $F_1F_2^2$ and F_1F_2 is obtained from the two initial blocks (00 11 22) and

(00 12 21). The six blocks are

(00 11 22)
(01 12 20)
(02 10 21)
(00 12 21)
(01 10 22)
(02 11 20)

This design belongs to the class of 2-cyclic designs defined in Section 4.5. It is also a partially balanced design with two associate classes (PBIB/2) since the 9 treatment combinations can be set out in the 3 × 3 array

00	01	02
10	11	12
20	21	22

with the property that two treatment combinations in the same row or column occur together in $\lambda_1 = 0$ blocks while two treatment combinations in different rows and columns occur together in $\lambda_2 = 1$ blocks. This array represents the association scheme of a Latin square type PBIB/2 design. The two distinct canonical efficiency factors of the design can be shown to be 1 and $\frac{1}{2}$, each with multiplicity 4. The first of these factors is associated with treatment contrasts corresponding to the two main effects, which are estimated with full information; while the second factor corresponds to interaction contrasts, which are estimable within blocks from half the observations.

In general, obtaining multi-replicate designs from the single replicate designs constructed by the method of Section 6.5 is too restrictive. Suppose, for instance, a design for a 3×2^2 experiment in blocks of 4 is required. The only single replicate design of this size would confound the main effect of the three-level factor, so that taking replications of this design would clearly be unsatisfactory. Again, for a 4^2 experiment in blocks of 6, no single replicate design can be constructed. The approach adopted in this chapter is to set out the treatment combinations of the factorial experiment in incomplete blocks using the designs of Chapters 3 and 4 and, in particular, to study those designs whose canonical efficiency factors can be identified with main effect and interaction treatment contrasts.

Designs based on single replicate designs are, as the above 3^2 example shows, included in this general framework.

The use of randomized block designs in factorial experiments has already been considered in Section 6.3. The analysis of such designs is straightforward as each treatment contrast is estimable entirely within blocks. Estimates are, therefore, based on unadjusted treatment means; see (6.14). These designs will not be discussed further here.

Results in Chapter 1 on the analysis of incomplete block designs and results on the treatment structure of factorial experiments given in Section 6.2 will be used. In particular, for some n-digit binary number $x = (x_1 x_2 \ldots x_n)$, contrasts in the $F_1^{x_1} F_2^{x_2} \ldots F_n^{x_n}$ interaction are estimated by

$$\mathbf{C}_x \hat{\boldsymbol{\tau}} = \mathbf{C}_x \boldsymbol{\Omega} \mathbf{q} \tag{7.1}$$

where $\mathbf{C}_x$ is given by (6.11), $\mathbf{q}$ is the vector of adjusted treatment totals given by (1.13) and $\boldsymbol{\Omega}$ is any generalized inverse of the information matrix $\mathbf{A}$ given by (1.12). Further,

$$\operatorname{cov}(\mathbf{C}_x \hat{\boldsymbol{\tau}}, \mathbf{C}_y \hat{\boldsymbol{\tau}}) = \mathbf{C}_x \boldsymbol{\Omega} \mathbf{C}_y \sigma^2 \tag{7.2}$$

7.2 Balanced incomplete blocks

Suppose that the v treatment combinations of a factorial experiment are set out in a balanced incomplete block design.

Example 7.1

A balanced incomplete block design for 4^2 treatment combinations in 16 blocks of 6 units per block is given in Table 7.1. It can be verified that each treatment combination is replicated six times and that every pair of treatment combinations occurs together in two blocks. The parameters of this design are, therefore, $v = 4^2 = 16$, $r = k = 6, b = 16$ and $\lambda = 2$.

No problems of analysis arise since it has already been shown in Section 3.3.2 that every treatment contrast is a basic contrast in a balanced incomplete block design, with all canonical efficiency factors equal to $E = \lambda v/rk$. The covariance matrix of the estimators of two different generalized interactions $\mathbf{C}_x \boldsymbol{\tau}$ and $\mathbf{C}_y \boldsymbol{\tau}$ $(x \neq y)$ is

$$\operatorname{cov}(\mathbf{C}_x \hat{\boldsymbol{\tau}}, \mathbf{C}_y \hat{\boldsymbol{\tau}}) = \mathbf{C}_x \boldsymbol{\Omega} \mathbf{C}_y \sigma^2 = (k/\lambda v) \mathbf{C}_x \mathbf{C}_y \sigma^2 = \mathbf{0} \tag{7.3}$$

Table 7.1 4^2 *experiment in 16 blocks of 6.*

(00	01	02	10	23	30)
(01	02	03	11	20	31)
(02	03	00	12	21	32)
(03	00	01	13	22	33)
(10	11	12	20	33	00)
(11	12	13	21	30	01)
(12	13	10	22	31	02)
(13	10	11	23	32	03)
(20	21	22	30	03	10)
(21	22	23	31	00	11)
(22	23	20	32	01	12)
(23	20	21	33	02	13)
(30	31	32	00	13	20)
(31	32	33	01	10	21)
(32	33	30	02	11	22)
(33	30	31	03	12	23)

since $\mathbf{\Omega} = (k/\lambda v)\mathbf{I}$ is a generalized inverse of the information matrix of a balanced incomplete block design, given in (3.5).

Hence, balanced incomplete block designs share with randomized block designs the important property that, for factorial experiments, the estimate of any contrast in one generalized interaction is orthogonal to the estimate of every contrast in any other generalized interaction. Every factorial effect can thus be estimated independently of other effects and the adjusted treatment sum of squares can consequently be partitioned orthogonally into sums of squares for each of the factorial effects.

In a balanced incomplete block design, the generalized interaction $\mathbf{C}_x\boldsymbol{\tau}$ is estimated by

$$\mathbf{C}_x\hat{\boldsymbol{\tau}} = (rE)^{-1}\mathbf{C}_x\mathbf{q} \tag{7.4}$$

i.e. by contrasts in the adjusted means of the treatment combinations. The variance–covariance matrix of this estimator is

$$V(\mathbf{C}_x\hat{\boldsymbol{\tau}}) = (rE)^{-1}\mathbf{C}_x\sigma^2 \tag{7.5}$$

and the sum of squares due to testing the hypothesis $H_0\!:\!\mathbf{C}_x\boldsymbol{\tau} = \mathbf{0}$ is

$$S(H_0) = (rE)^{-1}(\mathbf{C}_x\mathbf{q})'(\mathbf{C}_x\mathbf{q}) \tag{7.6}$$

These results are similar in form to those for randomized blocks given in Section 6.3. In particular, apart from the factor E^{-1}, the sum of squares in (7.6) is the same as that given in (6.16) but with unadjusted replaced by adjusted treatment means.

Since balanced incomplete block designs are efficiency-balanced, all main effect and interaction contrasts are estimated with the same precision. However, in a factorial experiment each treatment contrast need not be of equal importance. Interest may be primarily concerned with main effects and two-factor interactions, especially in experiments with many factors. High-order interactions may be of little importance or may even be assumed to be negligible. A design which provides more precise estimates of those contrasts of interest may therefore be preferred to a balanced incomplete block design. In any case, balanced incomplete block designs only exist for a limited number of parameter combinations, so that it will be necessary for this reason also to consider setting out factorial experiments in other incomplete block designs.

7.3 Group-divisible designs

For a group-divisible design, the $v = mn$ treatments are divided into m groups of n treatments each such that pairs of treatments belonging to the same group occur together in λ_1 blocks in the design, while pairs of treatments from different groups occur together in λ_2 blocks ($\lambda_1 \neq \lambda_2$).

Suppose a factorial experiment has $s + t$ factors with factors F_1, $F_2, \ldots, F_s$ at $m_1, m_2, \ldots, m_s$ levels respectively, and factors G_1, $G_2, \ldots, G_t$ at $n_1, n_2, \ldots, n_t$ levels respectively, such that $m = m_1 m_2 \ldots m_s$ and $n = n_1 n_2 \ldots n_t$. Let the different treatment groups in a group-divisible design correspond to different levels of the F factors and let treatments belonging to the same group correspond to different levels of the G factors. For instance, the correspondence between 12 treatments in 4 groups of 3 and the treatment combinations of a three-factor experiment with factors F_1, F_2 and G_1 at 2, 2 and 3 levels respectively is given in Table 7.2. Treatment number 8, for example, corresponds to factors F_1 and F_2 at levels 1 and 0 respectively and factor G_1 at level 2.

It will be convenient to distinguish between main effects and interactions involving F and G factors. Main effects will either be F-effects or G-effects. Interactions involving only the F factors, such as F_1F_2,

Table 7.2 *Treatment combinations of a $2^2 \times 3$ experiment in 4 groups of 3.*

	Levels of factor		*Levels of factor G_1*		
Group	F_1	F_2	*0*	*1*	*2*
1	0	0	0	1	2
2	0	1	3	4	5
3	1	0	6	7	8
4	1	1	9	10	11

will be called F-interactions, those involving only G factors will be called G-interactions, and interactions involving both F and G factors will be called FG-interactions.

It follows from (3.19) that contrasts in F-effects and F-interactions are basic contrasts with canonical efficiency factor $e_2 = mn\lambda_2/rk$. Similarly, from (3.18) contrasts in G-effects, G-interactions and FG-interactions are basic contrasts with canonical efficiency factor $e_1 = (rk - r + \lambda_1)/rk$.

Other results can be established from those of Section 3.5. For all $x \neq y \neq 0$, $\text{cov}(\mathbf{C}_x\hat{\boldsymbol{\tau}}, \mathbf{C}_y\hat{\boldsymbol{\tau}}) = \mathbf{0}$ so that all main effects and interactions are mutually orthogonal and can be estimated independently of each other. $\mathbf{C}_x\boldsymbol{\tau}$ is estimated by

$$\mathbf{C}_x\hat{\boldsymbol{\tau}} = (re_x)^{-1}\mathbf{C}_x\mathbf{q} \tag{7.7}$$

with

$$V(\mathbf{C}_x\hat{\boldsymbol{\tau}}) = (re_x)^{-1}\mathbf{C}_x\sigma^2 \tag{7.8}$$

where $e_x = e_2$ for F-effects and F-interactions, and $e_x = e_1$ otherwise. Apart from the efficiency factor, (7.7) and (7.8) are the same as the corresponding expressions for a balanced incomplete block design given by (7.4) and (7.5). Also, the sum of squares due to testing $H_0{:}\mathbf{C}_x\boldsymbol{\tau} = \mathbf{0}$ is

$$S(H_0) = (re_x)^{-1}(\mathbf{C}_x\mathbf{q})'(\mathbf{C}_x\mathbf{q}) \tag{7.9}$$

which again is the same form as (7.6). The analysis of a factorial experiment set out in a group-divisible design can, therefore, be seen to be relatively straightforward.

Since interactions are estimated with the same precision as either F-effects or G-effects, choosing a group-divisible design to provide good estimates of main effects will necessarily mean that equal importance will be attached to the high-order interactions. Hence, these designs, although simple to use and analyse, have the same drawbacks as balanced incomplete block designs when used in factorial experiments.

7.4 Factorial structure

If a factorial experiment is set out in a randomized block, balanced incomplete block or group-divisible design, then any treatment contrast belonging to one factorial effect (main effect or interaction) can be estimated independently of every contrast belonging to any other effect, i.e. factorial effects are mutually orthogonal. One consequence is that the adjusted treatment sum of squares can be partitioned orthogonally into sums of squares for each of the factorial effects. Hence, any effect can be estimated and assessed independently of any other effect. A further feature of such designs is that their canonical efficiency factors are directly associated with the different main effects and interactions, thereby permitting optimality criteria for choosing appropriate designs to be readily obtained. Incomplete block designs which possess these properties will be said to have (orthogonal) *factorial structure*. The problem of identifying which designs, or class of designs, have factorial structure will be considered in this section. It will be assumed that all designs are connected.

The v treatment combinations of an n-factor experiment with factor F_i at m_i levels $(i=1,2,\ldots,n)$ will be assumed to be in lexographical order, i.e. where the levels of the last factor are changed first. For instance, the 12 treatment combinations in the $2^2 \times 3$ experiment in Table 7.2 are set out in lexographical order. Now a design will have factorial structure, with respect to this ordering, if and only if treatment contrasts belonging to different factorial effects are uncorrelated, i.e. the covariance matrix given in (7.2) is zero. The adjusted treatment sum of squares $\hat{\boldsymbol{\tau}}'\mathbf{q}$ can then be partitoned orthogonally as follows:

$$\hat{\boldsymbol{\tau}}'\mathbf{q} = \sum_{x \neq 0} S(x)$$

where

$$S(x) = (\mathbf{C}_x\hat{\boldsymbol{\tau}})'(\mathbf{C}_x\boldsymbol{\Omega}\mathbf{C}_x')^{-}\mathbf{C}_x\hat{\boldsymbol{\tau}} \tag{7.10}$$

is the sum of squares due to testing $H_0\!:\mathbf{C}_x\boldsymbol{\tau}=\mathbf{0}$.

Mukerjee (1979) has established conditions for factorial structure in terms of the information matrix **A** of the design. First, define a proper matrix to be a square matrix with all row sums and all column sums equal. Now let the information matrix of the factorial design be given by

$$\mathbf{A} = \sum_{i=1}^{w} \xi_i(\mathbf{V}_{i1} \otimes \mathbf{V}_{i2} \otimes \cdots \otimes \mathbf{V}_{in}) \tag{7.11}$$

where w is a positive integer, $\xi_1, \xi_2, \ldots, \xi_w$ are real numbers and, for each i, $\mathbf{V}_{ij}$ is a proper matrix of order m_j $(j = 1, 2, \ldots, n)$. Mukerjee (1979) shows that a sufficient condition for a design to have factorial structure is that its information matrix be of the form given by (7.11). The condition is also shown to be necessary if the design is connected.

Sufficiency can be established as follows. With $\mathbf{C}_x$ given by (6.11) and **A** by (7.11), it follows that

$$\mathbf{C}_x\mathbf{A} = \mathbf{A}\mathbf{C}_x \tag{7.12}$$

since $\mathbf{VK} = \mathbf{KV}$ for any proper matrix **V**. Then, using (1.11),

$$\mathbf{C}_x\mathbf{A}\hat{\mathbf{t}} = \mathbf{A}\mathbf{C}_x\hat{\mathbf{t}} = \mathbf{C}_x\mathbf{q}$$

so that

$$\mathbf{C}_x\hat{\mathbf{t}} = \mathbf{\Omega}\mathbf{C}_x\mathbf{q}$$

where $\mathbf{\Omega}$ is any generalized inverse of **A**. Hence, as $V(\mathbf{q}) = \mathbf{A}\sigma^2$ and $\mathbf{C}_x\mathbf{C}_y = \mathbf{0}$ $(x \neq y)$ it is seen that

$$\operatorname{cov}(\mathbf{C}_x\hat{\mathbf{t}}, \mathbf{C}_y\hat{\mathbf{t}}) = \mathbf{\Omega}\mathbf{C}_x\mathbf{A}\mathbf{C}_y\mathbf{\Omega}'\sigma^2 = \mathbf{0}$$

for $x \neq y$. Thus a design whose information matrix **A** is given by (7.11) has factorial structure.

It can also be established that the canonical efficiency factors of a design with factorial structure can be identified with the main effects and interactions of the factorial experiment. Let **A** be written in canonical form as $\mathbf{A} = \sum \lambda_i \mathbf{p}_i \mathbf{p}_i'$ where $\mathbf{p}_i$ is a normalized eigenvector of **A** with non-zero eigenvalue λ_i. Then using (7.12),

$$\mathbf{C}_x\mathbf{A} = \sum \lambda_i (\mathbf{C}_x\mathbf{p}_i)(\mathbf{C}_x\mathbf{p}_i')$$

so that $\mathbf{C}_x\mathbf{A}$ can be written as

$$\mathbf{C}_x\mathbf{A} = \sum_i \lambda_{xi}\mathbf{t}_{xi}\mathbf{t}_{xi}' \tag{7.13}$$

where the $\mathbf{t}_{xi}$ are distinct vectors obtained from the $\mathbf{C}_x\mathbf{p}_i$ and normalized so that

$$\mathbf{t}'_{xi}\mathbf{t}_{xj} = \begin{cases} 1, & i = j \\ 0, & i \neq j \end{cases}$$

Again using (7.12),

$$\mathbf{C}_x\mathbf{A}\mathbf{C}_y\mathbf{A} = \sum_i\sum_j \lambda_{xi}\lambda_{yj}\mathbf{t}_{xi}\mathbf{t}'_{xi}\mathbf{t}_{yj}\mathbf{t}'_{yj} = \mathbf{0} \quad (x \neq y)$$

which implies that $\mathbf{t}_{xi} \neq \mathbf{t}_{yj}$ for all i and j. Hence

$$\mathbf{A} = \sum_x \mathbf{C}_x\mathbf{A} = \sum_x\sum_i \lambda_{xi}\mathbf{t}_{xi}\mathbf{t}'_{xi} \tag{7.14}$$

where the $\mathbf{t}_{xi}$ are distinct vectors for all x and i. Hence, $\mathbf{t}_{xi}$ is an eigenvector of $\mathbf{A}$ and is in the contrast space spanned by $\mathbf{C}_x$. The linear function $\mathbf{t}'_{xi}\boldsymbol{\tau}$ is, therefore, a basic contrast of the design and represents one degree of freedom of the generalized interaction $\mathbf{C}_x\boldsymbol{\tau}$ with corresponding canonical efficiency factor $e_{xi} = \lambda_{xi}/r$. The $v-1$ canonical efficiency factors are thus identified with the main effects and interactions of the factorial experiment. An average efficiency factor, E_x, can then be defined for the generalized interaction as

$$E_x = \nu_x / \sum e_{xi}^{-1} \tag{7.15}$$

where ν_x is the number of degrees of freedom of the interaction.

The generalized interaction $\mathbf{C}_x\boldsymbol{\tau}$ is estimated by $\mathbf{C}_x\hat{\boldsymbol{\tau}} = \mathbf{C}_x\boldsymbol{\Omega}\mathbf{q}$ with $V(\mathbf{C}_x\hat{\boldsymbol{\tau}}) = \mathbf{C}_x\boldsymbol{\Omega}\mathbf{C}_x\sigma^2$. The sum of squares (7.10) can be easily calculated. Using (7.12) and the estimability condition $\mathbf{C}_x = \mathbf{C}_x\boldsymbol{\Omega}\mathbf{A}$, it can be shown that $\mathbf{A}$ is a generalized inverse of $\mathbf{C}_x\boldsymbol{\Omega}\mathbf{C}_x$. Therefore, again using (7.12),

$$S(x) = \hat{\boldsymbol{\tau}}'\mathbf{C}_x\mathbf{A}\mathbf{C}_x\hat{\boldsymbol{\tau}} = \hat{\boldsymbol{\tau}}'\mathbf{C}_x\mathbf{A}\hat{\boldsymbol{\tau}}$$

so that, since $\mathbf{C}_x$ is idempotent and $\mathbf{A}\hat{\boldsymbol{\tau}} = \mathbf{q}$,

$$S(x) = (\mathbf{C}_x\hat{\boldsymbol{\tau}})'(\mathbf{C}_x\mathbf{q}) \tag{7.16}$$

Once the incomplete block analysis has been carried out, the further partitioning of the adjusted treatment sum of squares into components due to main effects and interactions is particularly straightforward. The formula given in (7.16) is similar in form to that given in (6.16) for randomized blocks, in (7.6) for balanced incomplete block designs and in (7.9) for group-divisible designs. Instead of calculations being based on sums of squares of unadjusted or adjusted treatment totals, (7.16) involves the sum of cross-products of adjusted treatment totals and estimates of treatment effects.

Many different classes of designs have factorial structure. They include randomized blocks, balanced incomplete blocks, group-divisible designs and Latin square type PBIB/2 designs; all these designs have information matrices which involve the Kronecker products of **I** and **J** matrices and, hence, satisfy the condition of (7.11). Also included are all designs satisfying property A of Kurkjian and Zelen (1963) and the n-cyclic designs of Section 4.5. Property A designs will be considered in the next section and n-cyclic designs in Section 7.6.

7.5 Balance

If a design has factorial structure then it has been shown that there exist $v-1$ linearly independent basic contrasts which span the contrast spaces defined by the main effects and interactions of the factorial experiment. The basic contrast $\mathbf{t}'_{xi}\boldsymbol{\tau}$ is estimated by

$$\mathbf{t}'_{xi}\hat{\boldsymbol{\tau}} = (re_{xi})^{-1}\mathbf{t}'_{xi}\mathbf{q}$$

with

$$\text{var}(\mathbf{t}'_{xi}\hat{\boldsymbol{\tau}}) = (re_{xi})^{-1}\sigma^2$$

Further, the sum of squares $S(x)$ given in (7.10) can be partitioned orthogonally into single degree of freedom components. Thus

$$S(x) = \sum_i (re_{xi})^{-1}(\mathbf{t}'_{xi}\mathbf{q})^2 \tag{7.17}$$

If the canonical efficiency factors e_{xi} corresponding to a particular generalized interaction are all distinct then this decomposition will be unique. At the other extreme, if all canonical efficiency factors are equal then any orthogonal decomposition is possible. A design is said to be *balanced* with respect to the generalized interaction $\mathbf{C}_x\boldsymbol{\tau}$ if and only if $e_{xi} = E_x$ for all i, where E_x is given in (7.15).

If a design is balanced with respect to all main effects and interactions, it will be said to have *factorial balance*. Such designs have a particularly simple analysis and also permit any single degree of freedom orthogonal decomposition of main effect and interaction sums of squares. Clearly, any design which is efficiency-balanced also has factorial balance. It follows that randomized blocks and balanced incomplete blocks have factorial balance. Group-divisible designs, on the other hand, are not efficiency-balanced but do have factorial balance, as was shown in Section 7.3. Kshirsagar (1966) has proved that a design has factorial balance if and only if it satisfies property A of Kurkjian and Zelen (1963), i.e. is a design whose information matrix is of the form

$$\mathbf{A} = \sum_{s} h(s_1, s_2, \ldots, s_n)(\mathbf{D}_{s_1} \otimes \mathbf{D}_{s_2} \otimes \cdots \otimes \mathbf{D}_{s_n}) \tag{7.18}$$

where each $s_j = 0$ or 1,

$$\mathbf{D}_{s_j} = \begin{cases} \mathbf{K}_{m_j}, & s_j = 1 \\ \mathbf{I}_{m_j}, & s_j = 0 \end{cases}$$

where the $h(s_1, s_2, \ldots, s_n)$ are constants depending on the s_j, and where the summation is over all binary numbers $s = (s_1 s_2 \ldots s_n)$. Such designs clearly have factorial structure since the information matrix has the form given by (7.11). It can be verified that the columns of $\mathbf{C}_x$ are eigenvectors of $\mathbf{A}$, so that the canonical efficiency factors are given by

$$E_x = (1/r)\sum h(s_1, s_2, \ldots, s_n) \tag{7.19}$$

where the summation now is over the subset of binary numbers $s = (s_1 s_2 \ldots s_n)$ defined by fixing $s_i = 0$ if $x_i = 1$ $(i = 1, 2, \ldots, n)$. For example, in a two-factor experiment $rE_{10} = h(0,0) + h(0,1)$ and $rE_{11} = h(0,0)$.

Included in this class are randomized blocks, balanced incomplete blocks, group-divisible and Latin square type PBIB/2 designs. The form of the $\mathbf{A}$ matrix in (7.18) can be used to check whether a design has factorial balance but it does not provide a general method of constructing such designs. Even when a factorial balanced design is given by a balanced incomplete block or group-divisible design it may be relatively inefficient in estimating the important factorial effects. The class of n-cyclic designs provides a simple method of constructing a large number of designs suitable for factorial experiments. Many of these designs have desirable properties with regard to both efficiency factors and balance. These designs will now be discussed.

7.6 n-Cyclic designs

In an n-cyclic design a treatment is represented by an n-tuple $a = a_1 a_2 \ldots a_n$ where $a_i = 0, 1, \ldots, m_i - 1$ $(i = 1, 2, \ldots, n)$. Designs are obtained by cyclical development of one or more initial blocks as described in Section 4.5. The information matrix of an n-cyclic design is, following (4.8),

$$\mathbf{A} = \sum_{h_1=0}^{m_1-1} \sum_{h_2=0}^{m_2-1} \cdots \sum_{h_n=0}^{m_n-1} \xi_{h_1h_2\ldots h_n}(\mathbf{\Gamma}_{h_1} \otimes \mathbf{\Gamma}_{h_2} \otimes \cdots \otimes \mathbf{\Gamma}_{h_n}) \qquad (7.20)$$

where $\xi_{h_1h_2\ldots h_n}$ are elements of the first row of $\mathbf{A}$ and where $\mathbf{\Gamma}_{h_i}$ is an $m_i \times m_i$ circulant matrix whose first row has 1 in the $(h_i + 1)$th column and zero elsewhere. Note that, since the eigenvectors of a circulant matrix do not depend on the individual elements of that matrix, there will exist a set of eigenvectors of the matrix $\mathbf{A}$ in (7.20) of the form

$$\gamma_1 \otimes \gamma_2 \otimes \cdots \otimes \gamma_n \qquad (7.21)$$

where γ_i is a vector of length m_i; see Section A.7 of the Appendix.

Now let the treatment $a = a_1a_2\ldots a_n$ correspond to a treatment combination in an n-factor experiment where a_i represents a level of factor F_i $(i = 1, 2, \ldots, n)$. It then follows that all n-cyclic designs have factorial structure, since the $\mathbf{A}$ matrix given in (7.20) has the same form as that given in (7.11). n-cyclic designs, therefore, provide a flexible class of incomplete block designs suitable for use in factorial experiments.

7.6.1 *Canonical efficiency factors*

For a factorial experiment of given size, a number of alternative n-cyclic designs will usually exist. The choice of an appropriate design can again be based on the canonical efficiency factors, and also on their degree of balance. The canonical efficiency factors are, from (4.9), given by

$$e_{u_1u_2\ldots u_n} = (1/r)\sum_{h_1}\sum_{h_2}\cdots\sum_{h_n} \xi_{h_1h_2\ldots h_n} \cos\left[\sum_{j=1}^{n} (2\pi u_j h_j/m_j)\right]$$

$$(u_l = 0, 1, \ldots, m_l - 1;\ l = 1, 2, \ldots, n) \qquad (7.22)$$

Excluded from (7.22) is the case where $u_l = 0$ for all l since $e_{00\ldots0} = 0$ is the eigenvalue of $\mathbf{A}$ corresponding to the vector $\mathbf{1}$.

With $u_l = 0$ for $l = 2, 3, \ldots, n$ and letting u_1 take non-zero values, (7.22) provides $m_1 - 1$ canonical efficiency factors given by

$$e_{u_10\ldots0} = (1/r)\sum_{h_1} \xi_{h_1} \cos(2\pi u_1 h_1/m_1) \qquad (7.23)$$

where

$$\xi_{h_1} = \sum_{h_2} \cdots \sum_{h_n} \xi_{h_1 h_2 \ldots h_n}$$

It can be seen, from a comparison with (4.2), that (7.23) also provides the canonical efficiency factors of a cyclic design whose information matrix is a circulant matrix with first-row elements given by ξ_{h_1} ($h_1 = 0, 1, \ldots, m_1 - 1$). Further, if γ_1 is an eigenvector of the information matrix of this cyclic design, then $\gamma_1 \otimes \mathbf{1} \otimes \cdots \otimes \mathbf{1}$ is an eigenvector of the $\mathbf{A}$ matrix given in (7.20). Hence, the canonical efficiency factors given by (7.23) are those associated with the main effect of factor F_1.

If $u_l = 0$ for $l = 3, 4, \ldots, n$ and if u_1 and u_2 take zero and non-zero values, other than $u_1 = u_2 = 0$, then (7.22) provides $m_1 m_2 - 1$ canonical efficiency factors given by

$$e_{u_1 u_2 0 \ldots 0} = (1/r) \sum_{h_1} \sum_{h_2} \xi_{h_1 h_2} \cos\left[(2\pi u_1 h_1/m_1) + (2\pi u_2 h_2/m_2)\right] \tag{7.24}$$

where

$$\xi_{h_1 h_2} = \sum_{h_3} \cdots \sum_{h_n} \xi_{h_1 h_2 \ldots h_n}$$

Again (7.24) provides the canonical efficiency factors of a 2-cyclic design whose information matrix has elements $\xi_{h_1 h_2}$. If $\gamma_1 \otimes \gamma_2$ is an eigenvector of this 2-cyclic design then $\gamma_1 \otimes \gamma_2 \otimes \mathbf{1} \otimes \cdots \otimes \mathbf{1}$ is an eigenvector of the $\mathbf{A}$ matrix given in (7.20). Hence, the canonical efficiency factors of (7.24) correspond to the F_1 main effect when $u_2 = 0$, the F_2 main effect when $u_1 = 0$, and to the $F_1 F_2$ interaction when both u_1 and u_2 are non-zero.

More generally, the canonical efficiency factors associated with the main effects and interactions of factors $F_1, F_2, \ldots, F_s$ ($s < n$) are obtained from (7.22) by putting $u_l = 0$ for $l = s + 1, \ldots, n$ and letting the other values be zero and non-zero. These factors will also be the canonical efficiency factors of an s-cyclic design whose information matrix has elements

$$\xi_{h_1 h_2 \ldots h_s} = \sum_{h_{s+1}} \cdots \sum_{h_n} \xi_{h_1 h_2 \ldots h_n} \tag{7.25}$$

The s-cyclic design whose information matrix has elements given by (7.25) can be obtained from the full n-cyclic design by first deleting from each treatment combination the last $n-s$ factors and then deleting any duplicate blocks from the resulting design. To show this, let $\mathbf{N}, \mathbf{N}_s$ and $\mathbf{A}, \mathbf{A}_s$ be the incidence and information matrices of the n-cyclic and derived s-cyclic designs respectively. Now $\mathbf{N}_s$ is obtained from $\mathbf{N}$ by summing the rows of $\mathbf{N}$ over the last $n-s$ factors, so that the relationship between $\mathbf{N}$ and $\mathbf{N}_s$ can be represented by

$$(\mathbf{I}\otimes\mathbf{K})\mathbf{N} = \mathbf{N}_s\otimes\mathbf{K} \tag{7.26}$$

where $\mathbf{I}$ and $\mathbf{K}$ are of order $m_1 m_2 \ldots m_s$ and $m_{s+1}\ldots m_n$ respectively. The presence of the $\mathbf{K}$ matrix on the right-hand side of (7.26) is due to duplicate blocks existing when factors are deleted from the n-cyclic design; these are eliminated in the s-cyclic design. It follows from (7.26) that

$$(\mathbf{I}\otimes\mathbf{K})\mathbf{A}(\mathbf{I}\otimes\mathbf{K}) = \mathbf{A}_s\otimes\mathbf{K}$$

and since, with $\mathbf{A}$ given by (7.20),

$$(\mathbf{I}\otimes\mathbf{K})\mathbf{A}(\mathbf{I}\otimes\mathbf{K}) = [\sum_{h_1}\cdots\sum_{h_s}\xi_{h_1\ldots h_s}(\mathbf{\Gamma}_{h_1}\otimes\cdots\otimes\mathbf{\Gamma}_{h_s})]\otimes\mathbf{K}$$

the result is established.

Hence, the canonical efficiency factors of the n-cyclic design corresponding to the main effects and interactions of factors $F_1, F_2, \ldots, F_s$ can be obtained either from (7.22) by putting $u_l = 0$ for $l = s+1, \ldots, n$, or as the canonical efficiency factors of the s-cyclic design with incidence matrix $\mathbf{N}_s$. The above results are set out in terms of the first s factors so as to avoid making the notation unnecessarily complicated, but these results apply equally well to any set of s factors.

Example 7.2
Consider the 3-cyclic design for a $4\times 3\times 2$ factorial experiment given by the initial block (000 001 221 310). Deleting, for instance, the second and third factors and eliminating duplicate blocks gives a cyclic design with 4 treatments having initial block (0 0 2 3). The canonical efficiency factors of this cyclic design are equal to those of the F_1 main effect in the three-factor design. Deleting the second factor only leads to a 4×2, 2-cyclic design with initial block (00 01 21 30). Its canonical efficiency factors are those of the F_1 and F_3 main effects and F_1F_3 interaction in the three-factor experiment.

In addition to the canonical efficiency factors, the degree of balance in an n-cyclic design can also be determined from the derived s-cyclic design. Since eigenvalues of $\mathbf{A}_s$ are also eigenvalues of $\mathbf{A}$, the n-cyclic design will be balanced with respect to all main effects and interactions involving the s factors if and only if the derived s-cyclic design has factorial balance.

Lewis and Dean (1985) have shown that this correspondence between the efficiency factors of the main effects and interactions of a subset s of the n factors and the efficiency factors of a derived s-factor design holds for all equally replicated designs with factorial structure.

7.6.2 Construction of n-cyclic factorial designs

It has been shown how the canonical efficiency factors of main effects, two-factor, three-factor,... interactions in an n-cyclic design can be directly obtained from cyclic, 2-cyclic, 3-cyclic,... designs. However, instead of breaking down the n-cyclic designs to obtain the canonical efficiency factors the reverse process can be used to build up the n-factor design. Cyclic designs can be merged to provide 2-cyclic designs; these in turn can be merged to give 3-cyclic designs; and so on.

Consider the main effect of factor F_1. Its $m_1 - 1$ canonical efficiency factors can be obtained directly from (7.23) or from the cyclic design whose initial blocks are obtained from the initial blocks of the n-cyclic design by deleting all factors except the first. All $m_1 - 1$ canonical efficiency factors are equal if the cyclic design is balanced. Further, its average efficiency factor $E(10\ldots0)$ is equal to the average efficiency factor E_1 of the cyclic design. The same results apply to any main effect. Hence optimal main effect n-cyclic designs are built up from optimal cyclic designs.

For the F_1F_2 interaction, attention is focused on the 2-cyclic design obtained from the n-cyclic design by deleting all factors except the first two. This 2-cyclic design has $m_1m_2 - 1$ canonical efficiency factors with $m_1 - 1$ and $m_2 - 1$ of them associated with the main effects of F_1 and F_2 respectively, and the remaining $(m_1 - 1)(m_2 - 1)$ factors associated with the F_1F_2 interaction. If they are all equal then the n-cyclic design is balanced with respect to this interaction. It follows that the cyclic designs chosen for the main effects of factors F_1 and F_2 should be merged so as to produce the most efficient

2-cyclic design if maximum precision is required for the F_1F_2 interaction. The following three examples illustrate the procedure.

Example 7.3
Consider the construction of a design for a 4^2 experiment in 16 blocks of 6 units per block. For both main effects, a cyclic design for 4 treatments in blocks of 6 is required. Two possible designs are given by initial blocks (0 0 0 1 2 3) and (0 0 1 1 2 3). The first design is balanced with average efficiency factor 0.89; the second design is not balanced but is A-optimal with average efficiency factor 0.96. A number of 2-cyclic designs can be obtained by merging these cyclic designs. One such design is given by the initial block (00 00 11 11 22 33); note that deleting either factor produces a cyclic design with initial block (0 0 1 1 2 3). Hence, this 2-cyclic design is obtained by using the second cyclic design above for both factors. Even though the average efficiency factors for both main effects are necessarily 0.96, the average efficiency factor for the interaction will be low, as this 2-cyclic design is relatively inefficient. A better 2-cyclic design based on the same cyclic designs is given by initial block (00 01 10 12 23 31); the average interaction efficiency factor being $E(11) = 0.831$. Although this design does not have factorial balance it is worth noting that in this case each factor could be represented by two pseudo-factors each at two levels and the design analysed as a factorial balanced 2^4 experiment.

As an alternative to the above, the 2-cyclic design with initial block (00 01 02 10 23 30) is a balanced incomplete block design and has been listed as such in Fisher and Yates (1963); there it is called a dicyclic design. The full design is, in fact, given in Table 7.1. It is based on the first of the 2-cyclic designs above and so all efficiency factors are 0.89. The other 2-cyclic design with initial block (00 01 10 12 23 31) may, however, be preferred if greater precision on main effects is required, at the expense of some loss in precision on the interaction. Such considerations become even more important as the number of factors increases, since the high-order interactions are rarely of much practical importance.

Example 7.4
Suppose a design is required for a 3×2^2 experiment in 12 blocks of 4 units per block with main effects and two-factor interactions

estimated with as high a degree of precision as possible. Firstly, for the main effects efficient cyclic designs for 3 and 2 treatments respectively in blocks of 4 are required. For 3 treatments the most efficient cyclic design has initial block (0 0 1 2); it is balanced with efficiency factor 0.94. For 2 treatments a fully efficient cyclic design is given by the initial block (0 0 1 1); it is a complete block design.

The next step is to merge these cyclic designs to produce 2-cyclic designs. For the F_1F_2 interaction a possible 3×2, 2-cyclic design, merging the cyclic designs (0 0 1 2) and (0 0 1 1), is given by the initial block (00 01 10 21); it has factorial balance. Another possible 2-cyclic design is given by the initial block (00 01 11 20); it is identical to the first one but will be used in the final 3-cyclic design for the F_1F_3 interaction. For the F_2F_3 interaction the 2^2, 2-cyclic design with initial block (00 01 10 11) is a complete block design and, hence, is fully efficient.

The final 3-cyclic design is obtained by merging the above cyclic and 2-cyclic designs. Any design obtained in this way will have high efficiency factors, and will be balanced, with respect to all main effects and two-factor interactions. Different choices of such 3-cyclic designs will only affect properties of the $F_1F_2F_3$ interaction. One possible design is obtained from the initial block (000 011 101 210) and can be shown to have factorial balance.

An alternative factorial balanced design is provided by a group-divisible design with $m = 6$ and $n = 2$; it is design R109 in the catalogue by Clatworthy (1973). The average efficiency factors for this group-divisible design and for the above 3-cyclic designs are

Table 7.3 *Efficiency factors for 3×2^2 designs.*

Generalized interaction	*3-cyclic design*	*Group-divisible design*
F_1	0.94	0.75
F_2	1.00	0.75
F_3	1.00	0.87
F_1F_2	0.81	0.75
F_1F_3	0.81	0.87
F_2F_3	1.00	0.87
$F_1F_2F_3$	0.44	0.87
Overall	0.75	0.81

given in Table 7.3. Although the group-divisible design has the higher overall efficiency factor, the 3-cyclic design may be preferred in a 3×2^2 factorial experiment since it is more efficient in estimating the main effects and two-factor interactions.

Example 7.5

n-Cyclic designs may be constructed when alternative balanced incomplete block and group-divisible (and other PBIB/2) designs are not available; the reverse will also be true. One such example is a 5×4 experiment in 20 blocks of 4. The 2-cyclic design with initial block (00 11 23 32) has factorial balance and is based on the A-optimal cyclic designs for both 5 and 4 treatments in blocks of 4.

Table 4.2 lists efficient binary cyclic designs for $4 \leqslant v \leqslant 15$ and these can be used in the first stages of the construction of *n*-cyclic designs aimed at maximizing main effect efficiency factors. As has been seen in the above examples, the cyclic designs needed for the *n*-cyclic designs are often non-binary. In such cases it is suggested that the binary designs are augmented by complete blocks. Certainly such designs are (M, S)-optimal. It has been conjectured by John and Williams (1982) that an A-optimal binary design augmented by complete blocks will produce A-optimal non-binary designs. In Example 7.3 the A-optimal binary cyclic design (0 1) for $k=2$ is augmented by a complete block design to give the A-optimal non-binary cyclic design (0 1 0 1 2 3) for $k=6$.

Bearing in mind the requirement for (M, S)-optimality, a general rule in constructing cyclic, 2-cyclic,... designs would be to ensure that treatment combinations occur as equally as possible in the initial block. In Example 7.3 the 2-cyclic design with initial block (00 00 11 11 22 33) is relatively inefficient because some treatment combinations occur twice, some once and others not at all in the initial block. A binary 2-cyclic design would be better in this case.

The designs in all three examples above have $b = v$ blocks. *n*-Cyclic designs with fewer blocks can be constructed if use is made of partial sets. The method of constructing such sets is given in Section 4.5. Basically the initial block will consist of a subgroup of the v treatment combinations of size d together with a number of its cosets, where d is a common divisor of v and k. The use of partial sets will, in general, impose further constraints on the efficiency factors of the main effects and interactions. For example, a 2-cyclic design for a

4^2 experiment in 8 blocks of 6 is given by the initial block (00 20 01 21 12 32). Compared with the design in 16 blocks, there is a decrease in the average efficiency factors on both main effects, with the decrease being greatest on the F_2 effect since the cyclic set (0 0 1 1 2 2) has had to be used.

Finally, a computer program can be easily written which will enable the properties of competing n-cyclic designs to be readily compared. The main steps of such a program are as follows:

1. Input the parameters and initial blocks of the n-cyclic design.
2. Obtain the first rows of the concurrence matrix $\mathbf{NN}'$ and the information matrix $\mathbf{A}$ from the initial blocks using a method of differencing similar to that described in Section 4.4 for cyclic designs.
3. Calculate the canonical efficiency factors, and average efficiency factors, of each main effect and interaction using (7.22).

7.6.3 Example of a two-factor experiment

The analysis of a factorial experiment involving two factors at 3 and 5 levels respectively in 15 blocks of 5 will now be given. The design used was a binary 2-cyclic design generated from the initial block (00 01 12 14 23). It can be shown to have factorial balance with efficiency factors

$$E(10) = 0.96, \qquad E(01) = 1.00, \qquad E(11) = 0.76$$

Although factorial balance enables considerable simplifications to be made in the analysis, the method of analysis given below is one that is appropriate for any design with factorial structure. In Section 8.6.2 the outline of a computer program is given which will provide an analysis of any block design; the use of such a program is the starting point for the analysis of factorial experiments.

The experiment could, of course, be analysed using analysis of variance or regression facilities available on many statistical computer packages; for instance, the experiment presented here is readily analysed using the ANOVA directive in the GENSTAT package.

The experiment was a pilot study for a larger study concerned with modelling annoyance due to combinations of noises; the results of the larger study are given in Rice and Izumi (1984). The noises were made up of three *sounds* each at five *intensity* levels. The sounds consisted of a single steady continuous sound, namely road traffic (R), and two intermittent discrete sounds, trains (T) and aircraft (A). The main purpose of the pilot study was to see if there were differences in annoyance between reactions to the continuous sound and the discrete sounds, with the aim of eliminating one of the discrete sounds from the larger study.

The second factor was the intensity of the sounds expressed in terms of the equivalent A-weighted sound pressure levels (L_{Aeq}). Five equally spaced levels were used, namely 35, 44, 53, 62 and 71 L_{Aeq}.

The experiment was carried out in a simulated domestic living-room listening facility at the Institute of Sound and Vibration Research, University of Southampton. The noises were played through multiple loudspeakers concealed behind the walls (for traffic and trains) and ceiling (for aircraft).

The fifteen noises were presented to fifteen subjects according to the design of Table 7.4; where, for example, A35 represents the aircraft sound at 35 L_{Aeq}. The duration of each noise was 10 minutes. During this time the road traffic sounds were heard continuously and the discrete sounds were heard at random intervals on 5 occasions, each discrete sound lasting about 30 seconds.

There was a short break between the presentation of each noise, during which the subjects reported their judged annoyance on a 0–9 numerical category scale, with the endpoints labelled 'not at all annoying' and 'extremely annoying'; these *subjective scale values* are the responses from the experiment.

The 2-cyclic design was set out in a row-by-column array with blocks corresponding to the columns of the array and with each row containing a single replicate of each treatment combination. The different levels of the sound and intensity factors were randomly allocated to the treatment combination labels, and subjects were randomly assigned to the columns of the array. The order of presentation of the treatment combinations to subjects was taken as a second blocking factor, and was randomly assigned to the rows of the array. The full design after randomization is given in Table 7.4. The subjective scale values together with subject and treatment combination totals are set out in Table 7.5.

The adjusted treatment totals $\mathbf{q}$ and the intra-block treatment estimates $\hat{\boldsymbol{\tau}}$ obtained from the analysis of the block design with 15 treatments in 15 blocks of 5 are set out in Tables 7.6 and 7.7 in the form of two-way tables classified according to the treatment factors;

Table 7.4 *Design for 3 × 5 experiment (after randomization).*

	Subject														
Order	*1*	*2*	*3*	*4*	*5*	*6*	*7*	*8*	*9*	*10*	*11*	*12*	*13*	*14*	*15*
1	A62	R62	A44	R53	T35	T62	R35	T71	A53	R71	T53	A35	A71	T44	R44
2	T44	A44	T53	A35	R53	R71	A71	R35	T35	A62	R44	T71	T62	R62	A53
3	R71	T35	R62	T62	A71	A44	T44	A62	R44	T53	A35	R53	R35	A53	T71
4	A53	R53	A35	R71	T44	T35	R62	T53	A71	R44	T62	A62	A44	T71	R35
5	R35	T71	R71	T44	A62	A53	T53	A44	R62	T35	A71	R44	R53	A35	T62

Table 7.5 *Subjective scale values.*

Treatment combination	*Subject*															*Totals*
	1	*2*	*3*	*4*	*5*	*6*	*7*	*8*	*9*	*10*	*11*	*12*	*13*	*14*	*15*	
A35			2	2							3	1		0		8
A44		2	3			4		3					4			16
A53	5					5			3					3	3	19
A62	5				6			5		5		3				24
A71					8		8		8		7		6			37
R35	1						2	2					4		1	10
R44									1	7	7	2			2	19
R53		6		6	7							6	5			30
R62		7	8				9		6					9		39
R71	9		9	8		9				9						44
T35		0			2	2			1	1						6
T44	4			0	2		4							2		12
T53			6				4	6		2	4					22
T62				7		7					5		7		5	31
T71		9						8				6		8	7	38
Totals	24	24	28	23	25	27	27	24	19	24	26	18	26	22	18	355

Table 7.6 *Adjusted treatment totals.*

	Intensity (L_{Aeq})					
Sound	*35*	*44*	*53*	*62*	*71*	*Total*
A	−15.4	−9.8	−3.0	1.0	12.4	−14.8
R	−13.8	−2.0	6.8	15.0	18.8	24.8
T	−17.8	−12.2	−3.8	7.0	16.8	−10.0
Total	−47.0	−24.0	0.0	23.0	48.0	0.0

Table 7.7 *Intra-block treatment estimates.*

	Intensity (L_{Aeq})					
Sound	*35*	*44*	*53*	*62*	*71*	*Total*
A	−2.90	−1.91	−0.63	−0.06	2.41	−3.08
R	−2.91	−0.29	1.52	3.19	3.66	5.17
T	−3.59	−2.60	−0.89	1.47	3.52	−2.08
Total	−9.40	−4.80	0.00	4.60	9.60	0.00

margin totals are also given. Note that margin intensity totals in Table 7.6 are $r = 5$ times those in Table 7.7; this being a consequence of the main effect of intensity being estimated with full efficiency, $E(01) = 1.00$. The intra-block analysis of variance is given in Table 7.8, where a sum of squares due to the order of presentation can be calculated directly from the order totals, since this second blocking factor is orthogonal to treatments.

Table 7.8 *Intra-block analysis of variance.*

Source of variation	*d.f.*	*s.s.*
Treatments (adjusted)	14	431.336
Subjects (unadjusted)	14	28.667
Order of presentation	4	10.267
Residual	42	58.396
Total	74	528.665

The adjusted treatment sum of squares can now be partitioned into orthogonal components representing the two main effects and the interaction. The calculations are, from (7.16), based on sums of cross-products of the various totals in Tables 7.6 and 7.7. They are:

$$S(\text{Sound}) = [(-14.8 \times -3.08) + (24.8 \times 5.17) + (-10.0 \times -2.08)]/5$$
$$= 38.920$$

$$S(\text{Intensity}) = [(-47 \times -9.4) + (-24 \times -4.8) + \cdots + (48 \times 9.6)]/3$$
$$= 374.533$$

$$S(\text{Sound by Intensity}) = 431.336 - 38.920 - 374.533 = 17.883$$

Since intensity is a quantitative factor, its main effect sum of squares can be further partitioned into linear, quadratic,... components. As the levels are equally spaced and the main effect is balanced, these components can be estimated and tested using tables of orthogonal polynomials (see Fisher and Yates, 1963; or John and Quenouille, 1977, p. 36). More generally, the components are estimated using regression techniques. For five equally spaced levels the vectors of linear and quadratic coefficients are $\mathbf{L}' = (-2\ -1\ 0\ 1\ 2)$ and $\mathbf{Q}' = (2\ -1\ -2\ -1\ 2)$ respectively. Note that $\mathbf{L}'\mathbf{L} = 10$ and $\mathbf{Q}'\mathbf{Q} = 14$. The two sums of squares, linear intensity (lin. int.) and quadratic intensity (quad. int.) are obtained from the quantities $\mathbf{L}'\mathbf{q}$, $\mathbf{L}'\hat{\mathbf{t}}$, $\mathbf{Q}'\mathbf{q}$ and $\mathbf{Q}'\hat{\mathbf{t}}$ calculated from the column totals in Tables 7.6 and 7.7; these quantities are given as the totals in Table 7.9. Hence,

$$S(\text{lin. int.}) = (237.0 \times 47.4)/(10 \times 3) = 374.46$$
$$S(\text{quad. int.}) = (3.0 \times 0.6)/(14 \times 3) = 0.043$$

In view of the structure of the sound factor, its main effect sum of squares can be partitioned into two orthogonal components

Table 7.9 *Basic calculations for linear and quadratic components.*

Sound	$\mathbf{L}'\mathbf{q}$	$\mathbf{L}'\hat{\mathbf{t}}$	$\mathbf{Q}'\mathbf{q}$	$\mathbf{Q}'\hat{\mathbf{t}}$
A	66.4	12.48	8.8	2.25
R	82.2	16.64	−16.6	−4.43
T	88.4	18.27	10.8	2.78
Total	237.0	47.40	3.0	0.60

representing the comparison between the continuous and discrete sounds and a comparison among the discrete sounds respectively. The appropriate vectors of coefficients are $\mathbf{C}' = (-1\ 2\ -1)$ and $\mathbf{D}' = (1\ 0\ -1)$, with $\mathbf{C}'\mathbf{C} = 6$ and $\mathbf{D}'\mathbf{D} = 2$. The continuous against discrete sound (CD sound) and the within discrete sound (AT sound) sums of squares are obtained from the quantities $\mathbf{C}'\mathbf{q}$, $\mathbf{C}'\hat{\mathbf{t}}$, $\mathbf{D}'\mathbf{q}$ and $\mathbf{D}'\hat{\mathbf{t}}$ calculated from the row totals in Tables 7.6 and 7.7. Hence

$$\begin{aligned} S(\text{CD sound}) &\\ &= \frac{[(2 \times 24.8) - (-14.8) - (-10.0)] \times [(2 \times 5.17) - (-3.08) - (-2.08)]}{6 \times 5} \\ &= 38.44 \end{aligned}$$

$$S(\text{AT sound}) = \frac{[-14.8 - (-10.0)] \times [-3.08 - (-2.08)]}{2 \times 5} = 0.48$$

Since the design has factorial balance, the interaction sum of squares can also be partitioned into sums of squares which represent interactions of the sound components with the intensity components. These sum of squares are obtained from the basic calculations given in Table 7.9. They are

$$\begin{aligned} S(\text{lin. int. by CD sound}) &\\ &= \frac{[(2 \times 82.2) - 66.4 - 88.4] \times [(2 \times 16.64) - 12.48 - 18.27]}{6 \times 10} \\ &= 0.41 \end{aligned}$$

$$\begin{aligned} S\ (\text{quad. int. by CD sound}) &\\ &= \frac{[(2 \times -16.6) - 8.8 - 10.8] \times [(2 \times -4.43) - 2.25 - 2.78]}{6 \times 14} \\ &= 8.73 \end{aligned}$$

$$S(\text{lin. int. by AT sound}) = \frac{(66.4 - 88.4) \times (12.48 - 18.27)}{2 \times 10} = 6.37$$

$$S(\text{quad. int. by AT sound}) = \frac{(8.8 - 10.8) \times (2.25 - 2.78)}{2 \times 14} = 0.04$$

The full analysis of variance is given in Table 7.10.

Table 7.10 *Full analysis of variance.*

Source of variation	*d.f.*		*s.s.*	*m.s.*	*F*
Intensity	4		374.533	93.63	67.3
lin. int.		1	374.46	374.46	269.3
quad. int.		1	0.04	0.04	0.0
remainder		2	0.03	0.02	0.0
Sound	2		38.920	19.46	14.0
CD sound		1	38.44	38.44	27.7
AT sound		1	0.48	0.48	0.4
Intensity by sound	8		17.883	2.24	1.6
lin. int. by CD sound		1	0.41	0.41	0.3
quad. int. by CD sound		1	8.73	8.73	6.3
lin. int. by AT sound		1	6.37	6.37	4.6
quad. int. by AT sound		1	0.04	0.04	0.0
remainder		4	2.34	0.59	0.4
Treatments (adjusted)	14		431.336	30.81	
Subjects (unadjusted)	14		28.667	2.05	
Order of presentation	4		10.267	2.57	
Residual	42		58.396	1.39	
Total	74		528.665		

A table of adjusted means can be obtained, following (1.18), by adding the overall mean of the responses, namely 4.73, to the estimates in Table 7.7; these means are given in Table 7.11. The standard errors of the differences (SED) between adjusted means can be obtained from the $\boldsymbol{\Omega}$ matrix of the incomplete block analysis.

Table 7.11 *Adjusted means.*

	Intensity (L_{Aeq})					
Sound	*35*	*44*	*53*	*62*	*71*	*Mean*
A	1.83	2.82	4.11	4.67	7.15	4.12
R	1.82	4.44	6.25	7.92	8.40	5.77
T	1.15	2.14	3.84	6.20	8.25	4.32
Mean	1.60	3.13	4.73	6.27	7.93	4.73

Mean standard errors can be obtained from the average efficiency factors; these standard errors will be exact for differences between main effect means since the design has factorial balance. The standard errors for means based on m observations are obtained from

$$\text{SED} = \sqrt{\left(\frac{2s^2}{mE}\right)}$$

where s^2 is the residual mean square and E the appropriate efficiency factor. For differences between sound and intensity adjusted means the standard errors are

$$\text{SED(Sound)} = \sqrt{\left(\frac{2 \times 1.39}{25 \times 0.96}\right)} = \pm 0.340$$

$$\text{SED(Intensity)} = \sqrt{\left(\frac{2 \times 1.39}{15 \times 1.0}\right)} = \pm 0.431$$

For differences between means in the main body of Table 7.11 the appropriate average efficiency factor is

$$E = \frac{14}{(2 \times .96^{-1}) + (4 \times 1) + (8 \times .76^{-1})} = 0.843$$

so that the average standard error is

$$\text{SED} = \sqrt{\left(\frac{2 \times 1.39}{5 \times 0.843}\right)} = \pm 0.812$$

The exact values are .821, .837 and .802 for differences between means from the same row, same column and different rows and columns respectively.

The conclusions from the analysis are that there are significant differences between intensity levels and between sounds. Further, differences between intensity levels are almost completely accounted for by the linear component. The estimate of the increase in subjective scale values per $9L_{\text{Aeq}}$ increase in intensity is

$$b = 47.4/(10 \times 3) = 1.58$$

There is also a significant difference between the continuous sound (road traffic) and the discrete sounds (aircraft and trains), but no difference between the two discrete sounds. Although the overall interaction mean square is not significant, there are two significant components. From Table 7.11, it is clear that one component is due to the fact that the largest differences between the discrete and continuous sounds are in the middle intensity levels, while the other component is an indication of the fact that the average scale values for trains, although starting lower, increase more rapidly with intensity than aircraft. The final conclusion from the pilot study was that there was little to choose between the two discrete sounds; in the event, aircraft was chosen for the larger study.

It may be appropriate in some experiments to recover the inter-block information available on treatment comparisons. The method of obtaining combined intra- and inter-block treatment estimates is described in Chapter 8. If a computer program is available which provides these estimates then the calculation of adjusted treatment means and standard errors for the factorial experiment follows in a similar way to that given above. In the experiment presented here, however, the recovery of inter-subject information is unlikely to be worthwhile, as there is little or no between-subject information available on the important main effects.

7.7 Other designs

If $\mathbf{N}_i$ is the incidence matrix of a block design for m_i treatments in blocks of size k_i with each treatment replicated r_i times $(i = 1, 2, \ldots, n)$, then

$$\mathbf{N} = \mathbf{N}_1 \otimes \mathbf{N}_2 \otimes \cdots \otimes \mathbf{N}_n \tag{7.27}$$

is the incidence matrix of an n-factor design for $v = m_1 m_2 \ldots m_n$ treatment combinations in blocks of size $k = k_1 k_2 \ldots k_n$ with each treatment combination replicated $r = r_1 r_2 \cdots r_n$ times. The designs have factorial structure and the canonical efficiency factors of any generalized interaction can be calculated from the designs obtained by deleting all factors not in the interaction. For instance, the efficiency factors of the main effect of factor F_1 are given by those of the design with incidence matix $\mathbf{N}_1$, while the efficiency factors of the main effects of F_1 and F_2 and of the $F_1 F_2$ interaction are given by the design with incidence matrix $\mathbf{N}_1 \otimes \mathbf{N}_2$. However, these n-factor designs have little practical value as the block size k or the number of replicates r or both will often be too large.

To overcome these difficulties, Mukerjee (1981) and Gupta (1983) have generalized the method to produce a number of series of designs with smaller block sizes and replications. Their methods produce very general families of designs, including for instance n-cyclic designs, which possess the same properties in terms of factorial structure and efficiency factors as the designs produced by (7.27). In general, however, the designs cannot be constructed in a simple systematic manner nor are explicit expressions available for the calculation of efficiency factors. Obtaining a factorial design of a given size with certain desirable properties could be a daunting task. Further research into the designs produced by these variants of the Kronecker product method could, however, be of considerable practical value.

7.8 Row–column designs

Factorial experiments can also be set out in $p \times q$ row–column designs as the example of Section 7.6.3 shows. The simplest type of design would be to use a Latin square for the v treatment combinations. The main effects and interactions would then be estimated independently of both row and column parameters. If insufficient experimental material was available to accommodate all treatment combinations in either the rows or the columns, then a row-orthogonal row–column design could be used; see Section 5.7. For instance, the 4^2 design in Table 7.1 has been set out in such a way that each of the 16 treatment combinations occurs once in each position within the blocks; the resulting design is thus a Youden square. The design in Table 7.4 is a row-orthogonal row–column design for a 3×5 experiment in 5 rows and 15 columns.

In factorial experiments, however, the number of treatment combinations v is often too large to use either a Latin square or a row-orthogonal design. It will frequently be necessary for both the number of rows and columns to be less than v. Certain treatment comparisons will, therefore, be confounded with the row and columns effects. The earliest type of row–column designs for factorial experiments were the quasi-Latin squares introduced by Yates (1937). Both Yates (1937) and Rao (1946) gave numerous designs, mainly for factors at two or three levels, in which certain main effects and interactions were either totally or partially confounded with rows or columns. The DSIGN algorithm given by Patterson (1976) is a general procedure for constructing factorial designs in various blocking structures including row–column designs. Examples of row–column factorial designs produced by this algorithm are given in Bailey (1977) and Patterson and Bailey (1978).

The n-cyclic designs defined in Section 4.5 have been shown to provide a flexible method of generating single, fractional and multiple replicate block designs for factorial experiments. John and Lewis (1983) have considered the use of such designs in row–column factorial experiments. Their method includes quasi-Latin squares and the row–column designs produced by the DSIGN algorithm if attention is again restricted to the subset of designs considered by Bailey (1977); see Section 6.11.

The method of construction ensures that the row–column design is such that both the row component design D_p and the column component design D_q are n-cyclic designs. Suppose the initial row consists of a subgroup G_1 of the treatment combinations of order d_1 together with a further $(q/d_1) - 1$ of its cosets. Since an n-cyclic set of v/d_1 blocks is generated from this row by the addition of one treatment combination from each of the cosets of G_1, the initial column must consist of pd_1/v elements from each of the cosets of G_1. A similar argument shows that the initial column must consist of a subgroup G_2 of order d_2 together with a further $(p/d_2) - 1$ of its cosets and that the initial row must consist of qd_2/v elements from each of the cosets of G_2. Clearly d_1 and d_2 must be common factors of v and q and v and p respectively, and v must be a divisor of pd_1 and qd_2. It can be easily verified that the resulting n-cyclic row–column design satisfies the adjusted orthogonality property given by (5.11), with $b = 1$, and has factorial structure.

Example 7.6
Consider the construction of a single replicate 3^4 experiment in a 9×9 row–column design. It is necessary to ensure that the initial

Table 7.12 3^4 *experiment in* 9×9 *row-column design.*

0000	1011	2022	1102	2110	0121	2201	0212	1220
1101	2112	0120	2200	0211	1222	0002	1010	2021
2202	0210	1221	0001	1012	2020	1100	2111	0122
0112	1120	2101	1211	2222	0200	2010	0021	1002
1210	2221	0202	2012	0020	1001	0111	1122	2100
2011	0022	1000	0110	1121	2102	1212	2220	0201
0221	1202	2210	1020	2001	0012	2122	0100	1111
1022	2000	0011	2121	0102	1110	0220	1201	2212
2120	0101	1112	0222	1200	2211	1021	2002	0010

row and column are chosen so that they only have treatment combination 0000 in common. This can be achieved, for instance, by taking the treatment combinations in the initial row to be given by the direct sum of the groups generated from 1011 and 1102, and in the initial column by the direct sum of groups generated from 1101 and 0112. The full design is given in Table 7.12. If the methods of Section 6.7 are applied separately to the row and column components, it can be shown that $F_1F_2^2F_3^2$, $F_1F_2F_4^2$, $F_1F_3F_4$ and $F_2F_3^2F_4$ are confounded with rows, and $F_1F_2^2F_3$, $F_1F_2F_4$, $F_1F_3^2F_4^2$ and $F_2F_3F_4^2$ with columns.

Example 7.7

John and Lewis (1983) give an example of a 3×4 experiment in a 4×6 row–column design. The initial row consists of subgroup (00 10 20) and coset (01 11 21), while the initial column consists of subgroup (00 02) and coset (11 13). The full design is:

00	01	10	11	20	21
02	03	12	13	22	23
11	12	21	22	01	02
13	10	23	20	03	00

It can be verified that each row has two treatments in common with every column, and thus satisfies the adjusted orthogonality condition. The canonical efficiency factors of the design can be obtained from (7.22) using the elements from the information matrix given by (5.3). Alternatively, the canonical efficiency factors can be obtained for the row and column components separately using (7.22) and then combined using (5.13). The efficiency factors for the F_1 and F_2 main effects are .75, .75 and .5, 1, .5 respectively, and for the F_1F_2 interaction 1, .25, 1, 1, .25, 1.

Further examples are given in John and Lewis (1983).

CHAPTER 8

Recovery of inter-block information

8.1 Introduction

The analysis of block designs presented in Chapter 1 was concerned with the estimation of treatment contrasts from comparisons made *within* blocks. If the blocks can be regarded as a random sample of blocks from some population then estimates of treatment contrasts may also be available from comparisons *between* blocks. Whether it is worth incorporating the information from this inter-block analysis with that from the within-block, or intra-block, analysis will depend on the extent to which the grouping of the experimental units into blocks has achieved a marked reduction in the error mean square. If this grouping is successful, i.e. block effects are large, then the amount of information which may be recovered from the inter-block analysis will be small. On the other hand, if block effects are small there may be substantial gains to be achieved by recovering the inter-block information.

Similarly, for row–column designs, there may be worthwhile gains to be achieved by recovering treatment information, where available, from comparisons made between rows and between columns. This chapter will be primarily concerned with the recovery of inter-block information from block designs and row–column designs. However, a general treatment of the subject will be outlined which should enable the methods of analysis to be applied to more complex blocking structures.

8.2 Orthogonal block structure

It is first necessary to consider the blocking structure of the experiment in order to determine what blocking comparison can be made, e.g. between blocks, within blocks, between columns, etc. The block structure derives from the use of plots which have an internal structure regardless of which treatments are applied to them.

Attention will be restricted to what Nelder (1965) calls *simple block structures*; the reasons for doing so will be discussed later in this section. For a block design this restriction means that it will be necessary for each block to contain the same number of plots, while in a $p \times q$ row–column design each row must contain q plots and each column p plots.

For a block design with b blocks each containing k plots, let y_{ij} represent the observation or yield on the ith plot of the jth block. Let $\bar{y}_j$ be the jth block mean and $\bar{y}$ the overall mean. Now consider the following *yield identity*:

$$y_{ij} = \bar{y} + (\bar{y}_j - \bar{y}) + (y_{ij} - \bar{y}_j)$$
$$(i = 1, 2, \ldots, k;\ j = 1, 2, \ldots, b) \tag{8.1}$$

The last two terms in (8.1) represent contrasts between blocks and within blocks respectively.

A matrix representation of this identity can be given by letting the observations be set out in a vector $\mathbf{y}$ such that the observations in the first k rows are from the first block, the observations in the next k rows are from the second block, and so on. Then (8.1) can be written as

$$\mathbf{y} = \mathbf{C}_0\mathbf{y} + \mathbf{C}_1\mathbf{y} + \mathbf{C}_2\mathbf{y} \tag{8.2}$$

where

$$\mathbf{C}_0 = \mathbf{K}_b \otimes \mathbf{K}_k$$
$$\mathbf{C}_1 = (\mathbf{I}_b - \mathbf{K}_b) \otimes \mathbf{K}_k$$
$$\mathbf{C}_2 = \mathbf{I}_b \otimes (\mathbf{I}_k - \mathbf{K}_k)$$

and where $\mathbf{K}_n$ is an $n \times n$ matrix with each element equal to $(1/n)$. Premultiplying a vector by $\mathbf{K}$ results in each element of the vector being replaced by its mean, whereas premultiplying by $\mathbf{I} - \mathbf{K}$ results in the mean being subtracted from each element in the vector. Hence, premultiplying $\mathbf{y}$ by $\mathbf{C}_0$ can be seen as an operation which first averages over plots and then over blocks to give a vector with each element equal to $\bar{y}$. Premultiplying $\mathbf{y}$ by $\mathbf{C}_1$ first averages over plots to give block means and then subtracts the overall mean from each block mean. Finally, premultiplying $\mathbf{y}$ by $\mathbf{C}_2$ results in block means being subtracted from the observations.

These $\mathbf{C}$ matrices form a *complete binary set* in that they are symmetric ($\mathbf{C}_i' = \mathbf{C}_i$), idempotent ($\mathbf{C}_i^2 = \mathbf{C}_i$), orthogonal ($\mathbf{C}_i\mathbf{C}_j = \mathbf{0}$, $i \neq j$) and satisfy $\mathbf{C}_0 + \mathbf{C}_1 + \mathbf{C}_2 = \mathbf{I}$.

A block structure which gives rise to the yield identity given in (8.1) or (8.2) is said to be *orthogonal.* The **C** matrices define the *strata* of this orthogonal block structure. The three strata are called respectively the *mean stratum*; the *block stratum*, representing between-block differences; and the *block by plot stratum*, representing between-plot, within-block differences.

The yield identity leads to a partitioning of the (uncorrected) total sum of squares into three components which correspond to the strata. From (8.1) the partition is

$$\sum\sum y_{ij}^2 = bk\bar{y}^2 + k\sum(\bar{y}_j - \bar{y})^2 + \sum\sum(y_{ij} - \bar{y}_j)^2 \tag{8.3}$$

or, in matrix notation using (8.2),

$$\mathbf{y}'\mathbf{y} = \mathbf{y}'\mathbf{C}_0\mathbf{y} + \mathbf{y}'\mathbf{C}_1\mathbf{y} + \mathbf{y}'\mathbf{C}_2\mathbf{y} \tag{8.4}$$

The degrees of freedom associated with the sum of squares $\mathbf{y}'\mathbf{C}_i\mathbf{y}$ is given by $\text{rank}(\mathbf{C}_i) = \text{trace}(\mathbf{C}_i)$, as $\mathbf{C}_i$ is idempotent. Thus, the total degrees of freedom $n = bk$ can be partitioned to give

$$bk = 1 + (b-1) + b(k-1) \tag{8.5}$$

These results are summarized in the *null* analysis of variance given in Table 8.1; it is called null since it is the analysis of variance that would be obtained if all plots received the same treatment. Usually the mean stratum is included in the total to give a (corrected) total sum of squares based on $bk - 1$ degrees of freedom. Note that this partition of degrees of freedom would then be the same as that given in Table 1.1.

The blocking structure of a row–column design can also be developed in a similar way. Let y_{ij} be the observation on the plot in the ith row and jth column. The appropriate yield identity now

Table 8.1 *Null analysis of variance for block designs.*

Stratum	*d.f.*	s.s.
Mean	1	$\mathbf{y}'\mathbf{C}_0\mathbf{y}$
Block	$b-1$	$\mathbf{y}'\mathbf{C}_1\mathbf{y}$
Block by plot	$b(k-1)$	$\mathbf{y}'\mathbf{C}_2\mathbf{y}$
Total	bk	$\mathbf{y}'\mathbf{y}$

becomes

$$y_{ij} = \bar{y} + (\bar{y}_{i.} - \bar{y}) + (\bar{y}_{.j} - \bar{y}) + (y_{ij} - \bar{y}_{i.} - \bar{y}_{.j} + \bar{y})$$
$$(i = 1, 2, \ldots, p; j = 1, 2, \ldots, q)$$

where $\bar{y}_{i.}$ and $\bar{y}_{.j}$ represent ith row and jth column means respectively. In matrix notation the identity is

$$\mathbf{y} = \mathbf{C}_0\mathbf{y} + \mathbf{C}_1\mathbf{y} + \mathbf{C}_2\mathbf{y} + \mathbf{C}_3\mathbf{y} \tag{8.6}$$

where now

$$\mathbf{C}_0 = \mathbf{K}_p \otimes \mathbf{K}_q$$
$$\mathbf{C}_1 = (\mathbf{I}_p - \mathbf{K}_p) \otimes \mathbf{K}_q$$
$$\mathbf{C}_2 = \mathbf{K}_p \otimes (\mathbf{I}_q - \mathbf{K}_q)$$
$$\mathbf{C}_3 = (\mathbf{I}_p - \mathbf{K}_p) \otimes (\mathbf{I}_q - \mathbf{K}_q)$$

It can be verified that these **C** matrices form a complete binary set.

The operations of premultiplying $\mathbf{y}$ by $\mathbf{C}_0, \mathbf{C}_1, \mathbf{C}_2$ and $\mathbf{C}_3$ are similar to those given above for block designs, where averaging and differencing are now carried out over rows and columns. Note also that these contrasts are identical in form to those used in (6.7) and (6.8) to define main effect and interaction contrasts in a two-factor experiment.

The four strata in a row–column design are respectively the mean stratum, the row stratum, the column stratum and the *row by column* stratum, involving interaction contrasts between rows and columns.

Sums of squares and degrees of freedom can be partitioned to give the null analysis of variance in Table 8.2.

A block design consists of two factors in which one factor (plots) is nested in the other (blocks). A row–column design has its two

Table 8.2 *Null analysis of variance for row–column designs.*

Stratum	*d.f.*	s.s.
Mean	1	$\mathbf{y}'\mathbf{C}_0\mathbf{y}$
Row	$p-1$	$\mathbf{y}'\mathbf{C}_1\mathbf{y}$
Column	$q-1$	$\mathbf{y}'\mathbf{C}_2\mathbf{y}$
Row by column	$(p-1)(q-1)$	$\mathbf{y}'\mathbf{C}_3\mathbf{y}$
Total	pq	$\mathbf{y}'\mathbf{y}$

factors rows and columns cross-classified. Many other blocking structures will have the experimental plots grouped into blocks in ways which can be described in terms of nesting and crossing. For instance, a nested row–column design, such as a lattice square, has rows crossed with columns which are then nested within blocks. For a simple orthogonal blocking structure the essential requirement is that the blocks all contain the same number of plots. Nelder (1965) gives simple rules for obtaining the yield identity and null analysis of variance for any such structure.

The importance of a simple block structure is that it has a complete randomization theory. The block structure is preserved under any permutation of the labelling of the plots. For instance, in a block design, the blocks may be labelled arbitrarily and the plots within blocks may also be labelled arbitrarily. There is no connection between the ith plot in the jth block and the ith plot in the j'th block. This arbitrary labelling can be achieved only if all blocks contain the same number of plots. If the suffices in $y_{ijk\ldots}$ identify the levels of the blocking factors then the randomization procedure for any simple block structure consists of reordering the values of each suffix at random.

The population of all possible vectors of observations generated by the randomization procedure gives the null randomization distribution; again assuming the null experiment in which all plots receive the same treatment. The variance–covariance matrix of this distribution, the null analysis of variance and expectations of sums of squares can be derived from the form of the simple block structure. Details of this randomization theory can be found in Nelder (1965).

8.3 Error covariance structure

Under the randomization procedure each observation will have the same variance and there will be a limited number of distinct covariances. For a block design there will be two covariances corresponding to whether any pair of observations are in the same block or in different blocks. For a row–column design there will be three covariances corresponding to pairs in the same row, same column or in different rows and columns.

First consider the block design. Following from (1.1), and ignoring treatment parameters as treatments play no part in the blocking structure, an appropriate model for a null block design is

$$y_{ij} = \mu + \beta_j + \varepsilon_{ij} \quad (i = 1, 2, \ldots, k; j = 1, 2, \ldots, b) \tag{8.7}$$

It will be assumed that the errors ε_{ij} are uncorrelated random variables with zero means and $\text{var}(\varepsilon_{ij}) = \sigma^2$, that the block effects β_j are also uncorrelated random variables with zero means and $\text{var}(\beta_j) = \sigma_1^2$ and that the β_j are uncorrelated with the ε_{ij}, i.e. $\text{cov}(\beta_j, \varepsilon_{ij}) = 0$. It then follows that

$$\text{var}(y_{ij}) = \sigma^2 + \sigma_1^2$$

$$\text{cov}(y_{ij}, y_{i'j'}) = \begin{cases} \sigma_1^2, & j = j', i \neq i' \\ 0, & j \neq j' \end{cases}$$

Thus two observations from the same block have covariance σ_1^2 whereas two observations from difference blocks have zero covariance.

Using matrix notation, the variance–covariance matrix of the y_{ij} is a block diagonal matrix which can be written as

$$V(\mathbf{y}) = \mathbf{V} = \mathbf{I}_b \otimes [(\sigma^2 + \sigma_1^2)\mathbf{I}_k + \sigma_1^2(\mathbf{J}_k - \mathbf{I}_k)]$$

i.e.

$$\mathbf{V} = \mathbf{I}_b \otimes [\sigma^2 \mathbf{I}_k + k\sigma_1^2 \mathbf{K}_k] \tag{8.8}$$

Now using the yield identity given in (8.2) it follows that

$$\mathbf{V} = (\mathbf{C}_0 + \mathbf{C}_1 + \mathbf{C}_2)'\mathbf{V}(\mathbf{C}_0 + \mathbf{C}_1 + \mathbf{C}_2)$$

and since it is easily shown that $\mathbf{C}_i\mathbf{V} = \xi_i \mathbf{C}_i$ $(i = 0, 1, 2)$ where

$$\xi_0 = \sigma^2 + k\sigma_1^2, \qquad \xi_1 = \sigma^2 + k\sigma_1^2, \qquad \xi_2 = \sigma^2 \tag{8.9}$$

then $\mathbf{V}$ can be written as

$$\mathbf{V} = \xi_0 \mathbf{C}_0 + \xi_1 \mathbf{C}_1 + \xi_2 \mathbf{C}_2 \tag{8.10}$$

Note that the columns of $\mathbf{C}_i$ are eigenvectors of $\mathbf{V}$ with eigenvalues ξ_i. Hence (8.10) represents a canonical or spectral decomposition of $\mathbf{V}$, which is then said to be in *spectral form.* Since the $\mathbf{C}_i$ matrices defined the strata of the blocking structure, the ξ_i are called the *stratum variances.*

A matrix representation of (8.7) is, following (1.3), given by

$$\mathbf{y} = \mathbf{1}\mu + \mathbf{Z}\boldsymbol{\beta} + \boldsymbol{\varepsilon}$$

where $\mathbf{Z}$ is the design matrix for blocks. With $V(\boldsymbol{\beta}) = \sigma_1^2\mathbf{I}$, $V(\boldsymbol{\varepsilon}) = \sigma^2\mathbf{I}$

and $\text{cov}(\boldsymbol{\beta}, \boldsymbol{\varepsilon}) = \mathbf{0}$,

$$\mathbf{V} = \sigma_1^2 \mathbf{Z}\mathbf{Z}' + \sigma^2 \mathbf{I} \tag{8.11}$$

which can be shown to be the same as (8.10) by noting that

$$\mathbf{Z}\mathbf{Z}' = k(\mathbf{I}_b \otimes \mathbf{K}_k) \tag{8.12}$$

For a row–column design, the null model is given by

$$y_{ij} = \mu + \rho_i + \gamma_j + \varepsilon_{ij} \tag{8.13}$$

where ρ_i and γ_j are row and column effects respectively. It is now assumed that the ρ_i, γ_j and ε_{ij} are random variables with zero means and with $\text{var}(\varepsilon_{ij}) = \sigma^2$, $\text{var}(\rho_i) = \sigma_1^2$ and $\text{var}(\gamma_j) = \sigma_2^2$. Further, all random variables are uncorrelated with each other. Then

$$\text{var}(y_{ij}) = \sigma^2 + \sigma_1^2 + \sigma_2^2$$

$$\text{cov}(y_{ij}, y_{i'j'}) = \begin{cases} \sigma_1^2, & i = i', j \neq j' \\ \sigma_2^2, & i \neq i', j = j' \\ 0, & i \neq i', j \neq j' \end{cases}$$

Thus the covariance between any two observations is σ_1^2, σ_2^2 or 0 depending on whether they are in the same row, same column or different rows and columns respectively.

In matrix notation, the variance–covariance matrix of y_{ij} is given by

$$\mathbf{V} = \mathbf{I}_p \otimes [(\sigma^2 + \sigma_1^2 + \sigma_2^2)\mathbf{I}_q + \sigma_1^2(\mathbf{J}_q - \mathbf{I}_q)] + (\mathbf{J}_p - \mathbf{I}_p) \otimes \sigma_2^2 \mathbf{I}_q$$

i.e.

$$\mathbf{V} = \mathbf{I}_p \otimes (\sigma^2 \mathbf{I}_q + q\sigma_1^2 \mathbf{K}_q) + p\sigma_2^2 \mathbf{K}_p \otimes \mathbf{I}_q$$

Using the $\mathbf{C}_i$ matrices given in (8.6) it can then be established that the spectral form of $\mathbf{V}$ is given by

$$\mathbf{V} = \xi_0 \mathbf{C}_0 + \xi_1 \mathbf{C}_1 + \xi_2 \mathbf{C}_2 + \xi_3 \mathbf{C}_3 \tag{8.14}$$

where the stratum variances are

$$\begin{aligned} \xi_0 &= \sigma^2 + q\sigma_1^2 + p\sigma_2^2 \\ \xi_1 &= \sigma^2 + q\sigma_1^2 \\ \xi_2 &= \sigma^2 + p\sigma_2^2 \\ \xi_3 &= \sigma^2 \end{aligned} \tag{8.15}$$

In summary, for any simple orthogonal block structure the *yield identity* can be represented by

$$\mathbf{y} = \sum \mathbf{C}_i \mathbf{y} \tag{8.16}$$

where the $\mathbf{C}_i$ matrices form a complete binary set and define the strata of the blocking structure. The null analysis of variance is given by the *quadratic identity*

$$\mathbf{y}'\mathbf{y} = \sum \mathbf{y}'\mathbf{C}_i \mathbf{y} \tag{8.17}$$

and the *error covariance structure* by

$$\mathbf{V} = \sum \xi_i \mathbf{C}_i \tag{8.18}$$

where the ξ_i are the stratum variances. Again simple rules for determining the ξ_i for different simple block structures can be found in Nelder (1965).

8.4 Generalized least squares analysis

Sections 8.2 and 8.3 were concerned with the null experiment in which all plots receive the same treatment. In the null model, therefore, each observation has the same expected value, namely μ. When different treatments, either as a single set of treatments or as combinations of a number of treatment factors, are applied to the plots, then the expected value of the observation from a particular plot will depend on which treatment is applied to that plot. Since the error covariance structure depends only on the blocking structure, the model is thus given by

$$\begin{aligned} E(\mathbf{y}) &= \mathbf{X}\boldsymbol{\tau} \\ V(\mathbf{y}) &= \mathbf{V} \end{aligned} \tag{8.19}$$

where $\mathbf{X}$ is the design matrix for treatments, $\boldsymbol{\tau}$ is the vector of treatment parameters and where $\mathbf{V}$ is given by (8.18).

Generalized least squares analysis can be used to provide an estimator $\tilde{\boldsymbol{\tau}}$ of $\boldsymbol{\tau}$. The estimator is obtained as a solution to the normal equation

$$\mathbf{X}'\mathbf{V}^{-1}\mathbf{X}\tilde{\boldsymbol{\tau}} = \mathbf{X}'\mathbf{V}^{-1}\mathbf{y}$$

Since $\mathbf{V}$ is in spectral form it follows that

$$\mathbf{V}^{-1} = \sum \xi_i^{-1} \mathbf{C}_i$$

Hence the normal equations become

$$\sum \xi_i^{-1}\mathbf{X}'\mathbf{C}_i\mathbf{X}\tilde{\boldsymbol{\tau}} = \sum \xi_i^{-1}\mathbf{X}'\mathbf{C}_i\mathbf{y} \tag{8.20}$$

The variance–covariance matrix of the estimator is then given by

$$V(\tilde{\boldsymbol{\tau}}) = \left(\sum \xi_i^{-1}\mathbf{X}'\mathbf{C}_i\mathbf{X}\right)^- \tag{8.21}$$

The matrix $\mathbf{A}_i = \mathbf{X}'\mathbf{C}_i\mathbf{X}$ is called the treatment *information matrix* in the ith stratum.

8.5 Estimation of stratum variances

If the stratum variances ξ_i are not known, as will usually be the case in practice, then estimates of them will be required in order to obtain $\tilde{\boldsymbol{\tau}}$ and $V(\tilde{\boldsymbol{\tau}})$ from (8.20) and (8.21) respectively. A number of methods of estimation have been given and three of them will be briefly discussed in this section.

In the intra-block analysis of variance given in Table 1.3 it was stated that the residual mean square s^2 provided an unbiased estimator of the variance σ^2. That is, the estimator of σ^2 is obtained by equating the residual mean square in the block by plot stratum to its expectation. More generally, an estimator of the ith stratum variance ξ_i can be obtained by equating the residual mean square in the ith stratum to its expectation. The expected value of this mean square will now be obtained.

The treatment information in the ith stratum will be obtained from comparisons made among the elements of the vector $\mathbf{z}_i = \mathbf{C}_i\mathbf{y}$. From (8.19), it follows that $E(\mathbf{z}_i) = \mathbf{C}_i\mathbf{X}\boldsymbol{\tau}$ and $V(\mathbf{z}_i) = \xi_i\mathbf{C}_i$, so that the least squares estimator $\tilde{\boldsymbol{\tau}}_i$ of $\boldsymbol{\tau}$ in the ith stratum is

$$\tilde{\boldsymbol{\tau}}_i = \mathbf{A}_i^-\mathbf{X}'\mathbf{C}_i\mathbf{y} \tag{8.22}$$

where $\mathbf{A}_i = \mathbf{X}'\mathbf{C}_i\mathbf{X}$. Hence the residual mean square in the ith stratum is given by

$$\begin{aligned}\mathrm{RSS}_i &= (\mathbf{z}_i - \mathbf{C}_i\mathbf{X}\tilde{\boldsymbol{\tau}}_i)'(\mathbf{z}_i - \mathbf{C}_i\mathbf{X}\tilde{\boldsymbol{\tau}}_i)\\ &= \mathbf{z}_i'(\mathbf{I} - \mathbf{X}\mathbf{A}_i^-\mathbf{X}')\mathbf{z}_i \end{aligned}\tag{8.23}$$

In order to obtain the expected value of RSS_i the following lemma will be needed.

Lemma 8.1

If $\mathbf{y}$ is a random vector distributed with mean vector $\boldsymbol{\mu}$ and variance–

covariance matrix $\mathbf{V}$ then the quadratic form $\mathbf{y}'\mathbf{A}\mathbf{y}$ has expectation

$$E(\mathbf{y}'\mathbf{A}\mathbf{y}) = \text{trace}(\mathbf{A}\mathbf{V}) + \boldsymbol{\mu}'\mathbf{A}\boldsymbol{\mu}$$

Proof By definition $\mathbf{V} = E[(\mathbf{y}-\boldsymbol{\mu})(\mathbf{y}-\boldsymbol{\mu})']$ so that $E(\mathbf{y}\mathbf{y}') = \mathbf{V} + \boldsymbol{\mu}\boldsymbol{\mu}'$. Since $\mathbf{y}'\mathbf{A}\mathbf{y} = \text{trace}(\mathbf{y}'\mathbf{A}\mathbf{y}) = \text{trace}(\mathbf{A}\mathbf{y}\mathbf{y}')$, then

$$E(\mathbf{y}'\mathbf{A}\mathbf{y}) = \text{trace}[\mathbf{A}(\mathbf{V}+\boldsymbol{\mu}\boldsymbol{\mu}')] = \text{trace}(\mathbf{A}\mathbf{V}) + \text{trace}(\boldsymbol{\mu}'\mathbf{A}\boldsymbol{\mu})$$

which proves the lemma.

Using Lemma 8.1,

$$\begin{aligned} E(\text{RSS}_i) &= \xi_i \text{trace}[(\mathbf{I}-\mathbf{X}\mathbf{A}_i^-\mathbf{X}')\mathbf{C}_i] + \boldsymbol{\tau}\mathbf{X}'\mathbf{C}_i(\mathbf{I}-\mathbf{X}\mathbf{A}_i^-\mathbf{X}')\mathbf{C}_i\mathbf{X}\boldsymbol{\tau} \\ &= \xi_i[\text{trace}(\mathbf{C}_i) - \text{trace}(\mathbf{X}'\mathbf{C}_i\mathbf{X}\mathbf{A}_i^-)] \\ &= \xi_i[\text{rank}(\mathbf{C}_i) - \text{rank}(\mathbf{A}_i)] \\ &= \xi_i d_i \end{aligned} \tag{8.24}$$

where d_i is therefore the degrees of freedom associated with RSS_i. Hence, the residual mean square RSS_i/d_i from the ith stratum gives an unbiased estimator of the stratum variance ξ_i.

The disadvantage of this method is that the residual mean square in the ith stratum may be based on an inadequate number of degrees of freedom, i.e. d_i can often be small or even zero. The reason for this is that although there may only be partial information on certain contrasts in the ith stratum, each independent contrast will account for a whole degree of freedom in calculating d_i. To overcome this difficulty Nelder (1968) advocates equating the *actual* residual mean square in the ith stratum to its expectation, i.e. using $\tilde{\boldsymbol{\tau}}$ from (8.20) instead of the ith stratum estimator $\tilde{\boldsymbol{\tau}}_i$ in (8.23). Both the residual mean square and its expectation will then involve the stratum variances so that an iterative procedure will be required to calculate the estimate of ξ_i. Nelder (1968) gives details of how this can be carried out for the class of *generally balanced* designs, i.e. designs for which the information matrices $\mathbf{A}_i = \mathbf{X}'\mathbf{C}_i\mathbf{X}$ are spanned by a common set of eigenvectors. Patterson and Thompson (1971) give a residual maximum likelihood method of estimating the stratum variances in block designs, which gives identical results to those of Nelder (1968) when block sizes are equal.

In recovering inter-block information from balanced incomplete block and lattice designs, Yates (1939, 1940b) estimated the stratum variances by equating the adjusted block mean square and residual mean square to their expectations, and showed that such a procedure resulted in little loss of efficiency. The method can be extended to block designs in general and to row–column designs with adjusted orthogonality. Since it provides simple useful estimates, this method will be used in subsequent sections.

It should be noted that if there is no treatment information available in the ith stratum then the combined estimate obtained from (8.20) will be equal to that obtained by setting $\xi_i^{-1} = 0$. For example, in a randomized block design all the treatment information is available in the block by plot stratum. Hence, the intra-block estimate $\hat{\tau}$ will be the same as the combined estimate $\tilde{\tau}$ no matter what estimate of ξ_1 is used. Finally, if it is unreasonable to consider the blocking effects in any stratum to be random variables, then the treatment information in the stratum concerned is ignored, by again setting the reciprocal of the stratum variance to zero.

8.6 Block designs

8.6.1 Combined treatment estimates

Let v treatments be set out in a block design consisting of b blocks each of k plots such that each treatment is replicated r times. The intra-block analysis has been considered by Chapter 1. To recover inter-block information it will be assumed that the block effects β_j and the error terms ε_{ij} are random variables with $E(\beta_j) = 0$, $\mathrm{var}(\beta_j) = \sigma_1^2$ and $E(\varepsilon_{ij}) = 0$, $\mathrm{var}(\varepsilon_{ij}) = \sigma^2$, and that all random variables are uncorrelated. The appropriate model for the combined analysis is then given by

$$E(\mathbf{y}) = \mathbf{X}\tau, \qquad V(\mathbf{y}) = \xi_0 \mathbf{C}_0 + \xi_1 \mathbf{C}_1 + \xi_2 \mathbf{C}_2 \qquad (8.25)$$

where $\mathbf{X}$ is the $n \times v$ design matrix for treatments, where the $\mathbf{C}_i$ matrices are defined in (8.6), and where the stratum variances are

$$\xi_0 = \xi_1 = \sigma^2 + k\sigma_1^2, \qquad \xi_2 = \sigma^2 \qquad (8.26)$$

The normal equations for the generalized least squares estimator of τ are then given by (8.20). The treatment information matrices are

$$\begin{aligned} \mathbf{A}_0 &= \mathbf{X}'\mathbf{C}_0\mathbf{X} = (r^2/n)\mathbf{J} \\ \mathbf{A}_1 &= \mathbf{X}'\mathbf{C}_1\mathbf{X} = (1/k)\mathbf{N}\mathbf{N}' - (r^2/n)\mathbf{J} \\ \mathbf{A}_2 &= \mathbf{X}'\mathbf{C}_2\mathbf{X} = r\mathbf{I} - (1/k)\mathbf{N}\mathbf{N}' \end{aligned} \qquad (8.27)$$

where use has been made of (1.4), (1.5) and the fact that, from (1.6) and (8.12),

$$\mathbf{NN'} = \mathbf{X'ZZ'X} = k\mathbf{X'}(\mathbf{I}_b \otimes \mathbf{K}_k)\mathbf{X}$$

Note that $\mathbf{A}_2$ is the information matrix $\mathbf{A}$ of the intra-block analysis given in (1.36). It is therefore the information matrix in the block by plot stratum.

Using (8.26) and (8.27),

$$\begin{aligned}\sum \xi_i^{-1}\mathbf{X'C}_i\mathbf{X} &= (\xi_0^{-1}/k)\mathbf{NN'} + \xi_2^{-1}[r\mathbf{I} - (1/k)\mathbf{NN'}] \\ &= \xi_2^{-1}\left\{r\mathbf{I} - (1/k)\left[1 - \frac{\xi_2}{\xi_0}\right]\mathbf{NN'}\right\}\end{aligned}$$

Now

$$1 - \frac{\xi_2}{\xi_0} = 1 - \frac{\sigma^2}{\sigma^2 + k\sigma_1^2} = \frac{k\sigma_1^2}{\sigma^2 + k\sigma_1^2} = \frac{1}{1 + \phi^{-1}}$$

where $\phi = k\sigma_1^2/\sigma^2$. Hence

$$\sum \xi_i^{-1}\mathbf{X'C}_i\mathbf{X} = \xi_2^{-1}\mathbf{A}_c$$

where

$$\mathbf{A}_c = r\mathbf{I} - (1/k^*)\mathbf{NN'} \tag{8.28}$$

and

$$k^* = k(1 + \phi^{-1}) \tag{8.29}$$

If $\mathbf{T}$ and $\mathbf{B}$ are the vectors of treatment and block totals respectively then, using (1.6) and (1.7), it can be seen that

$$\begin{aligned}\mathbf{q}_0 &= \mathbf{X'C}_0\mathbf{y} = r\bar{y}\mathbf{1} \\ \mathbf{q}_1 &= \mathbf{X'C}_1\mathbf{y} = (1/k)\mathbf{NB} - r\bar{y}\mathbf{1} \\ \mathbf{q}_2 &= \mathbf{X'C}_2\mathbf{y} = \mathbf{T} - (1/k)\mathbf{NB}\end{aligned} \tag{8.30}$$

Note again that $\mathbf{q}_2$ is the vector of adjusted treatment totals of the intra-block analysis given in (1.13). It follows that

$$\sum \xi_i^{-1}\mathbf{X'C}_i\mathbf{y} = (\xi_0^{-1}/k)\mathbf{NB} + \xi_2^{-1}[\mathbf{T} - (1/k)\mathbf{NB}]$$

i.e.

$$\sum \xi_i^{-1}\mathbf{X'C}_i\mathbf{y} = \xi_2^{-1}\mathbf{q}_c$$

where

$$\mathbf{q}_c = \mathbf{T} - (1/k^*)\mathbf{NB} \tag{8.31}$$

and k^* is given by (8.29). Hence, the generalized least squares normal equations are

$$\mathbf{A}_c\tilde{\boldsymbol{\tau}} = \mathbf{q}_c \tag{8.32}$$

where $\mathbf{A}_c$ and $\mathbf{q}_c$ are given by (8.28) and (8.31) respectively.

If $\phi^{-1} = 0$ then $k^* = k$ and the normal equations are those of the intra-block analysis given by (1.11). Thus, when the block variance σ_1^2 is very large compared with the error variance σ^2 there will be little to be gained from recovering inter-block information. The greatest gain is achieved when block differences are small.

In Section 1.3 the columns of the intra-block treatment information matrix $\mathbf{A}$ were taken to be spanned by a set of normalized eigenvectors $\mathbf{p}_1, \mathbf{p}_2, \ldots, \mathbf{p}_v$ with corresponding eigenvalues $\lambda_1, \lambda_2, \ldots, \lambda_v$. Since this means that

$$\mathbf{NN}'\mathbf{p}_i = k(r - \lambda_i)\mathbf{p}_i$$

it follows that $\mathbf{A}_c\mathbf{p}_i = \lambda_i^*\mathbf{p}_i$ where

$$\lambda_i^* = [r(k^* - k) + k\lambda_i]/k^* \tag{8.33}$$

Hence, unless $k = k^*$, all the eigenvalues of $\mathbf{A}_c$ are non-zero so that $\mathbf{A}_c$ is a non-singular matrix. The combined treatment estimator $\tilde{\boldsymbol{\tau}}$ is then given by

$$\tilde{\boldsymbol{\tau}} = \mathbf{A}_c^{-1}\mathbf{q}_c \tag{8.34}$$

with variance–covariance matrix

$$V(\tilde{\boldsymbol{\tau}}) = \mathbf{A}_c^{-1}\sigma^2 \tag{8.35}$$

Note that $\tilde{\boldsymbol{\tau}}'\mathbf{1} = v\bar{y}$ so that the mean of the combined treatment estimates is the overall mean $\bar{y}$, which means that they are directly comparable with the unadjusted treatment means and the intra-block estimates given by (1.18).

Estimates of σ^2 and σ_1^2 will be required to obtain $\tilde{\boldsymbol{\tau}}$. The residual mean square in the block by plot stratum gives an unbiased estimator of σ^2. The estimator of σ_1^2 will be obtained by equating the adjusted block sum of squares in the intra-block analysis to its expectation. This is the method of estimation used by Yates (1939, 1940b) for balanced incomplete block and lattice designs and used in the catalogues of cyclic designs by John, Wolock and David (1972) and partially balanced designs by Clatworthy (1973).

From (1.34), the adjusted block sum of squares is given by

$$S(\boldsymbol{\beta}/\mu, \boldsymbol{\tau}) = \mathbf{s}'\mathbf{L}^-\mathbf{s} \tag{8.36}$$

where

$$\mathbf{s} = \mathbf{B} - (1/r)\mathbf{N}'\mathbf{T}$$

and $\mathbf{L}^-$ is a generalized inverse matrix of

$$\mathbf{L} = k\mathbf{I} - (1/r)\mathbf{N}'\mathbf{N}$$

Now $\mathbf{s} = \mathbf{Z}'\mathbf{U}\mathbf{y}$ and $\mathbf{L} = \mathbf{Z}'\mathbf{U}\mathbf{Z}$, where $\mathbf{U} = \mathbf{I} - (1/r)\mathbf{X}\mathbf{X}'$ is an idempotent matrix, since $\mathbf{X}'\mathbf{X} = r\mathbf{I}$. Then

$$\mathbf{s}'\mathbf{L}^-\mathbf{s} = \mathbf{y}'\mathbf{U}\mathbf{Z}(\mathbf{Z}'\mathbf{U}\mathbf{Z})^-\mathbf{Z}'\mathbf{U}\mathbf{y}$$

so that, using (8.11) and Lemma 8.1,

$$E(\mathbf{s}'\mathbf{L}^-\mathbf{s}) = \text{trace}\,[\mathbf{U}\mathbf{Z}(\mathbf{Z}'\mathbf{U}\mathbf{Z})^-\mathbf{Z}'\mathbf{U}(\sigma^2\mathbf{I} + \sigma_1^2\mathbf{Z}\mathbf{Z}')]$$

since $\mathbf{X}'\mathbf{U} = \mathbf{0}$. Now

$$\text{trace}\,[\mathbf{U}\mathbf{Z}(\mathbf{Z}'\mathbf{U}\mathbf{Z})^-\mathbf{Z}'\mathbf{U}] = \text{trace}\,[(\mathbf{Z}'\mathbf{U}\mathbf{Z})^-\mathbf{Z}'\mathbf{U}\mathbf{Z}] = \text{rank}\,(\mathbf{L}) = b - 1$$

and

$$\text{trace}(\mathbf{U}\mathbf{Z}\mathbf{L}^-\mathbf{Z}'\mathbf{U}\mathbf{Z}\mathbf{Z}') = \text{trace}(\mathbf{L}\mathbf{L}^-\mathbf{L}) = \text{trace}(\mathbf{L}) = bk - v$$

Hence

$$E(\mathbf{s}'\mathbf{L}^-\mathbf{s}) = (b-1)\sigma^2 + (bk - v)\sigma_1^2$$

Let RMS and BMS represent the residual mean square and adjusted block mean square respectively; then it follows that estimators of σ^2 and σ_1^2 are given by

$$\hat{\sigma}^2 = \text{RMS}, \qquad \hat{\sigma}_1^2 = \frac{(b-1)(\text{BMS} - \text{RMS})}{bk - v} \tag{8.37}$$

Slightly different estimates are obtained for resolvable designs. Replicates are usually regarded as fixed, while blocks within replicates are random variables. Hence, the adjusted block sum of squares, BSS, for resolvable designs will be

$$\text{BSS} = \mathbf{s}'\mathbf{L}^-\mathbf{s} - (\mathbf{R}'\mathbf{R}/v - G^2/n) \tag{8.38}$$

where $\mathbf{R}$ is the vector of replication totals and G the overall total. In a similar way it can be shown that

$$E(\text{BSS}) = (b - r)\sigma^2 + (bk - v - rk + k)\sigma_1^2$$

so that

$$\hat{\sigma}^2 = \text{RMS}, \qquad \hat{\sigma}_1^2 = \frac{(b-r)(\text{BMS} - \text{RMS})}{bk - v - rk + k} \tag{8.39}$$

Standard errors of differences between adjusted treatment means are obtained from the variance–covariance matrix given in (8.35), where variance components σ^2 and σ_1^2 are replaced by their estimates. Following (2.2), the average standard error is given by

$$\text{ASE} = \sqrt{(2\hat{\sigma}^2/rE^*)} \tag{8.40}$$

where

$$E^* = [(k^* - k) + kE]/k^*$$

is given by replacing each λ_i in (8.33) by rE, where E is the average (harmonic mean) efficiency factor.

8.6.2 Analysis of block designs by computer

The analysis of any block design, with or without the recovery of inter-block information, can be easily carried out on a computer. The extensive and flexible matrix operation facilities in GENSTAT, for example, make the writing of a general block design program in this statistical language a relatively simple task; similar facilities are also available in the statistical computer package SAS. The programming language APL is also ideally suited for such a program, since it readily handles matrix operations. Writing programs in languages such as FORTRAN and PASCAL is also relatively straightforward if access is available to libraries of scientific subroutines.

In order to enable both resolvable and non-resolvable block designs to be analysed in the same program, it will be assumed that the experimental area is divided into r' superblocks, with each superblock divided into b/r' blocks and with each block divided into k plots. A resolvable design has $r' = r$ so that the superblocks correspond to the replication groups, whereas in a non-resolvable design $r' = 1$; more generally designs which have every treatment replicated r/r' times in each superblock could be analysed.

Assuming a connected design, the main steps of a block design program are as follows:

1. Input the parameters v, b, r, r' and k.
2. Since each observation is uniquely identified with a treatment, superblock and block, the observations together with associated levels of the treatment, superblock and block factors are input. If the levels of the superblocks and blocks are in lexographical

order then they can, alternatively, be set up within the program rather than input separately.

3. From the arrangement of treatments in blocks the incidence matrix **N** is calculated. Also the vectors **T**, **B** and **R** of treatment block and superblock totals respectively are calculated.
4. Proceed with the intra-block analysis. First calculate the information matrix **A** given in (1.12) and the vector of adjusted treatment totals **q** given in (1.13). Next calculate a generalized inverse of **A** as $\mathbf{\Omega} = (\mathbf{A} + \mathbf{J})^{-1}$ and obtain the treatment estimates $\hat{\boldsymbol{\tau}}$ and the adjusted treatment means $\hat{\boldsymbol{\tau}}^*$ given in (1.15) and (1.18) respectively. Finally, calculate the analysis of variance as given in Table 1.3. Standard errors of differences between treatment estimates can be obtained from $\mathbf{\Omega}\hat{\sigma}^2$, where $\hat{\sigma}^2$ is the residual mean square in the analysis of variance. An average standard error is given by $\sqrt{(2\hat{\sigma}^2/rE)}$, where the average efficiency factor E can be calculated by

$$E = \frac{v-1}{r[\text{trace}(\mathbf{\Omega}) - (1/v)]}$$

5. Calculate the residuals. The estimates of the block parameters $\boldsymbol{\beta}$ are obtained from (1.14) with $\hat{\mu} = \bar{y}$. For observation y_{ij} corresponding to treatment i in the jth block, the residual is then given by $r_{ij} = y_{ij} - \hat{y}_{ij}$, where

$$\hat{y}_{ij} = \bar{y} + \hat{\tau}_i + \hat{\beta}_j$$

6. Recover inter-block information, if required. The estimate of σ_1^2 is obtained from (8.39) with r replaced by r'. Note that the adjusted block sum of squares is given by (8.38), where $\mathbf{s}'\mathbf{L}^-\mathbf{s} = S(\boldsymbol{\beta}/\mu, \tau)$ is obtained using (1.33). The combined treatment estimates are given by (8.34) where

$$\mathbf{A}_c = f\mathbf{A} + r(1-f)\mathbf{I}$$
$$\mathbf{q}_c = f\mathbf{q} + (1-f)\mathbf{T}$$

and where

$$f = k\hat{\sigma}_1^2/(\hat{\sigma}^2 + k\hat{\sigma}_1^2)$$

Again standard errors of the differences between the combined treatment estimates can be obtained from $\mathbf{A}_c^{-1}\hat{\sigma}^2$. Alternatively, an average standard error is given by (8.40) where

$$E^* = 1 - f(1-E)$$

8.6.3 *Example of the analysis of a block design*

The results of a spring oats trial grown in Craibstone, near Aberdeen, have been kindly made available by the North of Scotland College of Agriculture. The trial involved 24 varieties and three replicates, each consisting of six blocks of four plots. Thus the design parameters are $v = 24$, $r = r' = 3$, $b = 18$ and $k = 4$. The resolvable block design used was an $\alpha(0, 1)$-design obtained from the α-array:

$$\begin{matrix} 0 & 0 & 0 \\ 0 & 1 & 5 \\ 0 & 3 & 2 \\ 0 & 2 & 3 \end{matrix}$$

This array is obtained from Table 4.5 by taking the first three rows and first four columns of the array with $s = k = 6$. The α-design has an average efficiency factor of $E = 0.7265$.

The varieties were labelled $1, 2, \ldots, 24$ and were randomly assigned to the treatment labels in the α-design. Replicates, blocks within replicates and plots within blocks were also randomly permuted. The design after randomization is given in Table 8.3. The plots were, in fact, laid out in a single line of 72, with variety 11 on the 1st plot, 4 on the 2nd, 5 on the 3rd, ..., and 7 on the 72nd. The dry matter grain yields (in tonne/ha) are given in Table 8.4, where each yield can be identified with a variety according to the design in Table 8.3. For instance, the yield of variety 14 on plot 2 of block 3 in replicate 1 is 4.7572.

The intra-block analysis of variance is given in Table 8.5. A comparison of the adjusted varieties mean square with the residual

Table 8.3 *Resolvable design for variety trial (after randomization).*

Replicate 1						*Replicate 2*						*Replicate 3*					
Block						*Block*						*Block*					
1	*2*	*3*	*4*	*5*	*6*	*7*	*8*	*9*	*10*	*11*	*12*	*13*	*14*	*15*	*16*	*17*	*18*
11	21	23	13	17	6	8	24	12	5	2	19	11	2	17	12	21	3
4	10	14	3	15	12	20	15	11	9	18	7	1	15	18	13	22	5
5	20	16	19	7	24	14	3	21	10	13	6	14	9	4	10	16	20
22	2	18	8	1	9	4	23	17	1	22	16	19	8	6	23	24	7

Table 8.4 *Dry matter grain yield (tonne/ha).*

Replicate 1					
1	*2*	*3*	*4*	*5*	*6*
4.1172	4.6540	4.2323	4.2530	4.7876	4.7085
4.4461	4.1736	4.7572	3.3420	5.0902	5.2560
5.8757	4.0141	4.4906	4.7269	4.1505	4.9577
4.5784	4.3350	3.9737	4.9989	5.1202	3.3986

Replicate 2					
7	*8*	*9*	*10*	*11*	*12*
3.9926	3.9039	5.3127	5.1202	5.1566	5.3148
3.6056	4.9114	5.1163	4.2955	5.0988	4.6297
4.5294	3.7999	5.3802	4.9057	5.4840	5.1751
4.3599	4.3042	5.0744	5.7161	5.0969	5.3024

Replicate 3					
13	*14*	*15*	*16*	*17*	*18*
3.9205	4.0510	4.3234	4.1746	4.4130	2.8873
4.6512	4.6783	4.2486	4.7512	4.2397	4.1972
4.3887	3.1407	4.3960	4.0875	4.3852	3.7349
4.5552	3.9821	4.2474	3.8721	3.5655	3.6096

mean square gives an F value of 5.25 on 23 and 31 degrees of freedom, which is highly significant. Thus there are significant differences between varieties. The adjusted intra-block variety means, together with the average standard error of differences between adjusted means (SED), are given in Table 8.6. The plot of residuals against fitted values did not indicate any obvious departures from the assumptions underlying the analysis.

The estimates of σ^2 and σ_1^2 are $\hat{\sigma}^2 = 0.0833$ and $\hat{\sigma}_1^2 = 0.0588$ respectively. The adjusted combined variety means are also given

Table 8.5 *Analysis of variance.*

Source of variation	*d.f.*	*s.s.*	*m.s.*
Replicates	2	6.1362	3.0681
Blocks within replicates (unadj.)	15	7.6204	0.5080
Varieties (adj.)	23	10.0619	0.4375
Blocks within replicates (adj.)	15	3.6019	0.2401
Varieties (unadj.)	23	14.0803	0.6122
Residual	31	2.5819	0.0833
Total	71	26.4004	

in Table 8.6; the combined estimates have been arranged in order of magnitude to facilitate the comparison of varieties. The average standard error of differences is 0.264, corresponding to a value of $E^* = 0.798$. The exact standard errors range from 0.254 to 0.267.

The gain in *efficiency* resulting from dividing the 24 plots within each replicate into six blocks of four can be calculated as follows. If blocks within replicates are ignored, the residual mean square from the resulting complete block analysis would be given by pooling the adjusted block mean square of 0.2401 and the residual mean square of 0.0833. This gives a complete block residual mean square of

$$\frac{15 \times 0.2401 + 31 \times 0.0833}{46} = 0.1344$$

so that the standard error of mean differences would have been $\sqrt{(2 \times 0.1344/3)} = 0.299$; compared with 0.264 for the combined analysis of the α-design. The efficiency is then

$$\text{Efficiency} = \left(\frac{0.299}{0.264}\right)^2 = 1.286$$

Thus there has been a 28.6% increase in efficiency resulting from the use of blocks within replicates.

The Craibstone trial was one of many similar trials covering different parts of Scotland and a range of seasonal conditions. The

Table 8.6 *Adjusted variety means.*

Variety	*Intra-block estimates*	*Combined estimates*
1	5.08	5.11
5	5.03	5.04
15	5.02	4.97
19	4.84	4.84
21	4.76	4.80
14	4.90	4.77
13	4.73	4.76
12	4.64	4.76
16	4.72	4.73
17	4.51	4.60
6	4.43	4.54
22	4.46	4.53
8	4.67	4.53
4	4.54	4.49
2	4.47	4.48
10	4.36	4.37
18	4.32	4.36
11	4.22	4.28
23	4.31	4.25
24	4.14	4.15
7	4.11	4.11
20	4.20	4.04
9	3.44	3.50
3	3.61	3.50
SED	0.276	0.264

results of this analysis (adjusted variety means and standard error) went forward to be coordinated with the results of other trials.

8.7 Combined analysis for row–column designs

For v treatments set out in a $p \times q$ row–column design in which each treatment is replicated r times, the appropriate model for the combined analysis is

$$E(\mathbf{y}) = \mathbf{X}\boldsymbol{\tau}, \qquad V(\mathbf{y}) = \xi_0\mathbf{C}_0 + \xi_1\mathbf{C}_1 + \xi_2\mathbf{C}_2 + \xi_3\mathbf{C}_3 \qquad (8.41)$$

where the $\mathbf{C}_i$ matrices are given in (8.6) and the stratum variances ξ_i in (8.15). The information matrices required in the normal equations (8.20) are

$$\begin{aligned}\mathbf{A}_0 &= \mathbf{X}'\mathbf{C}_0\mathbf{X} = (r^2/n)\mathbf{J}\\ \mathbf{A}_1 &= \mathbf{X}'\mathbf{C}_1\mathbf{X} = (1/q)\mathbf{N}_p\mathbf{N}_p' - (r^2/n)\mathbf{J}\\ \mathbf{A}_2 &= \mathbf{X}'\mathbf{C}_2\mathbf{X} = (1/p)\mathbf{N}_q\mathbf{N}_q' - (r^2/n)\mathbf{J}\\ \mathbf{A}_3 &= \mathbf{X}'\mathbf{C}_3\mathbf{X} = r\mathbf{I} - (1/q)\mathbf{N}_p\mathbf{N}_p' - (1/p)\mathbf{N}_q\mathbf{N}_q' + (r^2/n)\mathbf{J}\end{aligned} \qquad (8.42)$$

where $\mathbf{N}_p$ and $\mathbf{N}_q$ are the $v \times p$ row-treatment and $v \times q$ column-treatment incidence matrices respectively. Note that

$$\mathbf{A}_3 = \mathbf{A}_p + \mathbf{A}_q - r\mathbf{I} + (r^2/n)\mathbf{J}$$

where, following from results in Section 5.2, $\mathbf{A}_p$ and $\mathbf{A}_q$ are the information matrices in respectively the row by plot stratum of the row component design D_p and the column by plot stratum of the column component design D_q. Hence $\mathbf{A}_3$ is equal to the information matrix $\mathbf{A}$ given in (5.3) for the fixed effects model where, for $b = 1$, $\mathbf{A}_b = r\mathbf{I} - (r^2/n)\mathbf{J}$. Also, from a comparison with $\mathbf{A}_1$ in (8.27), it can be seen that $\mathbf{A}_1$ and $\mathbf{A}_2$ in (8.42) are the information matrices in the between-row stratum of D_p and the between-column stratum of D_q respectively.

Analogous to (8.28) and (8.29) for the combined analysis of block designs, define

$$\mathbf{A}_{pc} = r\mathbf{I} - (1/q^*)\mathbf{N}_p\mathbf{N}_p' \qquad (8.43)$$

where $q^* = q(1 + \phi_1^{-1})$ and $\phi_1 = q\sigma_1^2/\sigma^2$ for the row component design D_p, and

$$\mathbf{A}_{qc} = r\mathbf{I} - (1/p^*)\mathbf{N}_q\mathbf{N}_q' \qquad (8.44)$$

where $p^* = p(1 + \phi_2^{-1})$ and $\phi_2 = p\sigma_2^2/\sigma^2$ for the column component design D_q. Then, using (8.15) and (8.42) and after some simplification,

$$\sum \xi_i^{-1}\mathbf{X}'\mathbf{C}_i\mathbf{X} = \xi_3^{-1}\mathbf{A}_c$$

where

$$\mathbf{A}_c = \mathbf{A}_{pc} + \mathbf{A}_{qc} - r\mathbf{I} + (r^2/n)\xi\mathbf{J} \qquad (8.45)$$

and where

$$\xi = \xi_3(\xi_0^{-1} - \xi_1^{-1} - \xi_2^{-1} + \xi_3^{-1})$$

If $\mathbf{T}$, $\mathbf{R}$ and $\mathbf{C}$ are the vectors of treatment, row and column totals respectively, then it can be shown that

$$\begin{aligned}
\mathbf{q}_0 &= \mathbf{X}'\mathbf{C}_0\mathbf{y} = r\bar{y}\mathbf{1} \\
\mathbf{q}_1 &= \mathbf{X}'\mathbf{C}_1\mathbf{y} = (1/q)\mathbf{N}_p\mathbf{R} - r\bar{y}\mathbf{1} \\
\mathbf{q}_2 &= \mathbf{X}'\mathbf{C}_2\mathbf{y} = (1/p)\mathbf{N}_q\mathbf{C} - r\bar{y}\mathbf{1} \\
\mathbf{q}_3 &= \mathbf{X}'\mathbf{C}_3\mathbf{y} = \mathbf{T} - (1/q)\mathbf{N}_p\mathbf{R} - (1/p)\mathbf{N}_q\mathbf{C} + r\bar{y}\mathbf{1}
\end{aligned} \tag{8.46}$$

Again $\mathbf{q}_3$ is the vector of adjusted treatment totals given in (5.4) for the fixed effects model. Analogous to (8.31) define

$$\begin{aligned}
\mathbf{q}_{pc} &= \mathbf{T} - (1/q^*)\mathbf{N}_p\mathbf{R} \\
\mathbf{q}_{qc} &= \mathbf{T} - (1/p^*)\mathbf{N}_q\mathbf{C}
\end{aligned}$$

Then

$$\sum \xi_i^{-1}\mathbf{X}'\mathbf{C}_i\mathbf{y} = \xi_3^{-1}\mathbf{q}_c$$

where

$$\mathbf{q}_c = \mathbf{q}_{pc} + \mathbf{q}_{qc} - \mathbf{T} + r\bar{y}\xi\mathbf{1} \tag{8.47}$$

Hence the generalized least squares normal equations are

$$\mathbf{A}_c\tilde{\boldsymbol{\tau}} = \mathbf{q}_c \tag{8.48}$$

where $\mathbf{A}_c$ and $\mathbf{q}_c$ are given by (8.45) and (8.47) respectively.

If both $\phi_1^{-1} = 0$ and $\phi_2^{-1} = 0$ the normal equations are the same as those of the fixed effects analysis given in (5.2). When both ϕ_1^{-1} and ϕ_2^{-1} are zero, the matrix $\mathbf{A}_c$ is non-singular and the combined treatment estimator is then given by

$$\tilde{\boldsymbol{\tau}} = \mathbf{A}_c^{-1}\mathbf{q}_c \tag{8.49}$$

with variance–covariance matrix

$$V(\tilde{\boldsymbol{\tau}}) = \mathbf{A}_c^{-1}\sigma^2 \tag{8.50}$$

An unbiased estimator of σ^2 is provided by the residual mean square in the row by column stratum. However, the problem of estimating the variance components σ_1^2 and σ_2^2 in a row–column design is, in general, a difficult one; see for example Roy and Shah (1961), Shah (1962, 1977) and Wheeler and Kshirsagar (1981). The iterative method given by Nelder (1968) is available for generally balanced row–column designs, i.e. designs whose information matrices given in (8.42) are spanned by a common set of eigenvectors. An important subclass of generally balanced designs are those that satisfy the adjusted orthogonality condition of (5.1), with $b = 1$. Eccleston and John (1986) show that for adjusted orthogonal designs, estimates of σ_1^2 and σ_2^2 can be simply obtained by equating adjusted row and column sums of squares to their expectations, i.e. using the same method of estimation for the row and column component designs as used for block designs. This follows as a direct consequence of the orthogonality between rows and columns after adjusting for treatments.

Consider the fixed-effects row–column model where $\boldsymbol{\tau}$, $\boldsymbol{\rho}$ and $\boldsymbol{\gamma}$ are the vectors of treatment, row and column parameters respectively, i.e. model (5.1) with $b = 1$. Eliminating the treatment parameters from the full set of normal equations gives the equations

$$(q\mathbf{I} - \mathbf{N}_p'\mathbf{N}_p/r)\hat{\boldsymbol{\rho}} + (\mathbf{J} - \mathbf{N}_p'\mathbf{N}_q/r)\hat{\boldsymbol{\gamma}} = \mathbf{R} - (1/r)\mathbf{N}_p'\mathbf{T}$$
$$(\mathbf{J} - \mathbf{N}_q'\mathbf{N}_p/r)\hat{\boldsymbol{\rho}} + (p\mathbf{I} - \mathbf{N}_q'\mathbf{N}_q/r)\hat{\boldsymbol{\gamma}} = \mathbf{C} - (1/r)\mathbf{N}_q'\mathbf{T}$$

With the adjusted orthogonality condition $\mathbf{N}_p'\mathbf{N}_q = r\mathbf{J}$ it can be seen that $\mathbf{L}\hat{\boldsymbol{\rho}} = \mathbf{s}$, where

$$\mathbf{L} = q\mathbf{I} - (1/r)\mathbf{N}_p'\mathbf{N}_p, \qquad \mathbf{s} = \mathbf{R} - (1/r)\mathbf{N}_p'\mathbf{T}$$

The adjusted row sum of squares is then

$$S(\boldsymbol{\rho}/\mu, \boldsymbol{\tau}, \boldsymbol{\gamma}) = \mathbf{s}'\mathbf{L}^-\mathbf{s}$$

which is of the same form as the adjusted block sum of squares given in (8.36). It can thus be shown that

$$E(\mathbf{s}'\mathbf{L}^-\mathbf{s}) = (p-1)\sigma^2 + (pq - v)\sigma_1^2$$

The adjusted column sum of squares is calculated in the same way and has expectation $(q-1)\sigma^2 + (pq - v)\sigma_2^2$. Thus estimates of σ_1^2 and σ_2^2 can be obtained by equating the adjusted row and column sums of squares to their expectations.

The importance of adjusted orthogonality in the construction of row–column designs has been emphasized in Sections 5.4 and 5.8. These designs now have the further advantage that, when treatment information can be obtained from at least two of the blocking strata, simple estimates of the variance components are available.

APPENDIX

Some matrix results

The main matrix results required for this book are given in this Appendix. Most of them are stated without proof as they are relatively straightforward and can be found in most books on matrix algebra; see for instance Searle (1966), Basilevsky (1983) and Graybill (1983). The properties of circulant matrices are considered in more detail as they have particular relevance in the book and are perhaps not widely known. A detailed account of circulant matrices can be found in Davis (1979).

A.1 Notation

An $n \times m$ matrix $\mathbf{A} = ((a_{ij}))$ has n rows and m columns with a_{ij} being the element in the ith row and jth column. $\mathbf{A}'$ will denote the transpose of $\mathbf{A}$. $\mathbf{A}$ is *symmetric* if $\mathbf{A} = \mathbf{A}'$. If $m = 1$ then the matrix will be called a *column vector* and will usually be denoted by a lower case letter, $\mathbf{a}$ say. The transpose of $\mathbf{a}$ will be a *row vector* $\mathbf{a}'$.

A diagonal matrix whose diagonal elements are those of the vector $\mathbf{a}$, and whose off-diagonal elements are zero, will be denoted by $\mathbf{a}^{\delta}$. The inverse of $\mathbf{a}^{\delta}$ will be denoted by $\mathbf{a}^{-\delta}$. More generally, if each element in $\mathbf{a}$ is raised to power p then the corresponding diagonal matrix will be denoted by $\mathbf{a}^{p\delta}$.

Some special matrices and vectors are:

$\mathbf{1}_n$,	the $n \times 1$ vector with every element unity
$\mathbf{I}_n = \mathbf{1}_n^{\delta}$,	the $n \times n$ identity matrix
$\mathbf{J}_{n \times m} = \mathbf{1}_n \mathbf{1}'_m$,	the $n \times m$ matrix with every element unity
$\mathbf{J}_n = \mathbf{J}_{n \times n}$	
$\mathbf{K}_n = (1/n)\mathbf{J}_n$	the $n \times n$ matrix with every element n^{-1}.

Suffices can be omitted if, in doing so, no ambiguity results.

A.2 Trace and rank

If $\mathbf{A} = ((a_{ij}))$ is a square matrix of order n, i.e. an $n \times n$ matrix, then

$$\text{trace}(\mathbf{A}) = \sum_{i=1}^{n} a_{ii}$$

The number of linearly independent rows or columns of the $n \times n$ matrix $\mathbf{A}$ is given by $r = \text{rank}(\mathbf{A})$. If $r = n$ then $\mathbf{A}$ is non-singular, if $r < n$ $\mathbf{A}$ is singular.

Provided the matrices are conformable, and not necessarily square, then

$$\text{trace}(\mathbf{AB}) = \text{trace}(\mathbf{BA}) \tag{A.1}$$

$$\text{rank}(\mathbf{AB}) \leqslant \min[\text{rank}(\mathbf{A}), \text{rank}(\mathbf{B})] \tag{A.2}$$

$$\text{rank}(\mathbf{AA}') = \text{rank}(\mathbf{A}'\mathbf{A}) = \text{rank}(\mathbf{A}) = \text{rank}(\mathbf{A}') \tag{A.3}$$

It follows from (A.1) that the trace of the product of matrices is invariant under any cyclic permutation of the matrices. For example,

$$\text{trace}(\mathbf{ABC}) = \text{trace}(\mathbf{BCA}) = \text{trace}(\mathbf{CAB})$$

A.3 Eigenvalues and eigenvectors

Let $\mathbf{A}$ be a square symmetric matrix of order n. An *eigenvalue* of $\mathbf{A}$ is a scalar λ such that $\mathbf{Ax} = \lambda\mathbf{x}$ for some vector $\mathbf{x} \neq \mathbf{0}$. The vector $\mathbf{x}$ is called an *eigenvector* of $\mathbf{A}$.

If $\lambda_1, \lambda_2, \ldots, \lambda_n$ are the eigenvalues of $\mathbf{A}$ then

$$\text{trace}(\mathbf{A}) = \sum_{i=1}^{n} \lambda_i \tag{A.4}$$

$$\text{rank}(\mathbf{A}) = \text{number of non-zero eigenvalues} \tag{A.5}$$

$$|\mathbf{A}| = \prod_{i=1}^{n} \lambda_i \tag{A.6}$$

where $|\mathbf{A}|$ is the determinant of $\mathbf{A}$. If $\lambda_1, \lambda_2, \ldots, \lambda_m$ are the *distinct* eigenvalues of $\mathbf{A}$ $(m \leqslant n)$ then the determinant can be written as

$$|\mathbf{A}| = \lambda_1^{n_1} \lambda_2^{n_2} \ldots \lambda_m^{n_m}$$

where n_i is the *multiplicity* of λ_i, and where $\sum n_i = n$.

Repeated application of $\mathbf{Ax} = \lambda\mathbf{x}$ shows that, for some positive integer h, $\mathbf{x}$ is also an eigenvector of $\mathbf{A}^h$ with corresponding eigenvalue

λ^h, i.e.

$$\mathbf{A}^h\mathbf{x} = \lambda^h\mathbf{x} \qquad (h = 1, 2, \ldots) \tag{A.7}$$

The eigenvalues of a symmetric matrix are real. For each eigenvalue of a symmetric matrix there exists a real eigenvector.

If $\mathbf{B}$ is a non-square matrix then $\mathbf{B}'\mathbf{B}$ and $\mathbf{B}\mathbf{B}'$ have the same non-zero eigenvalues.

For every symmetric matrix $\mathbf{A}$ there exists a non-singular matrix $\mathbf{X}$ such that

$$\mathbf{X}^{-1}\mathbf{A}\mathbf{X} = \boldsymbol{\lambda}^{\delta} \tag{A.8}$$

where the elements of $\boldsymbol{\lambda}$ are the eigenvalues of $\mathbf{A}$.

For the symmetric matrix $\mathbf{A}$ there exists a set of orthogonal and normalized eigenvectors $\mathbf{x}_1, \mathbf{x}_2, \ldots, \mathbf{x}_n$ satisfying

$$\mathbf{x}_i'\mathbf{x}_j = \begin{cases} 1, & i = j \\ 0, & i \neq j \end{cases}$$

The *canonical* or *spectral* decomposition of $\mathbf{A}$ is then given by

$$\mathbf{A} = \sum_{i=1}^{n} \lambda_i \mathbf{x}_i \mathbf{x}_i' \tag{A.9}$$

where λ_i is the eigenvalue corresponding to $\mathbf{x}_i$. If $\mathbf{A}$ is non-singular then

$$\mathbf{A}^{-1} = \sum_{i=1}^{n} \lambda_i^{-1} \mathbf{x}_i \mathbf{x}_i' \tag{A.10}$$

It also follows from (A.9) that the eigenvectors $\mathbf{x}_i$ $(i = 1, 2, \ldots, n)$ satisfy

$$\sum_{i=1}^{n} \mathbf{x}_i \mathbf{x}_i' = \mathbf{I} \tag{A.11}$$

A.4 Idempotent matrices

A square matrix $\mathbf{A}$ is *idempotent* if $\mathbf{A}^2 = \mathbf{A}$. If $\mathbf{A}$ is idempotent then so is $\mathbf{I} - \mathbf{A}$. $\mathbf{I}$ and $\mathbf{K}$ are examples of idempotent matrices.

The eigenvalues of an idempotent matrix $\mathbf{A}$ of order n are 1 with multiplicity r, and 0 with multiplicity $n - r$, where $r = \text{rank}(\mathbf{A})$. It then follows from (A.4) and (A.5) that for an idempotent matrix

$$\text{rank}(\mathbf{A}) = \text{trace}(\mathbf{A}) \tag{A.12}$$

If $\mathbf{A}$ and $\mathbf{B}$ are idempotent then $\mathbf{AB}$ is idempotent if and only if $\mathbf{AB} = \mathbf{BA}$.

A.5 Generalized inverses

Let $\mathbf{A}$ be a square matrix of order n with rank$(\mathbf{A}) = r$. If $r = n$ then $\mathbf{A}$ has an unique inverse $\mathbf{A}^{-1}$ which satisfies

$$\mathbf{A}\mathbf{A}^{-1} = \mathbf{A}^{-1}\mathbf{A} = \mathbf{I} \tag{A.13}$$

Further, an unique solution to the equations $\mathbf{Ax} = \mathbf{y}$ is then given by

$$\mathbf{x} = \mathbf{A}^{-1}\mathbf{y} \tag{A.14}$$

If $r < n$, $\mathbf{A}^{-1}$ does not exist. There will, however, be an infinite number of solutions of the consistent equations $\mathbf{Ax} = \mathbf{y}$. As an example, consider the equations

$$\begin{pmatrix} 1 & 2 \\ 2 & 4 \end{pmatrix}\begin{pmatrix} x_1 \\ x_2 \end{pmatrix} = \begin{pmatrix} 3 \\ 6 \end{pmatrix}$$

The second equation is twice the first so that rank$(\mathbf{A}) = 1$. Replacing 6 by 5, say, in the vector $\mathbf{y}$ would lead to an inconsistent set of equations for which no solution will exist.

One solution to the above equations is given by $x_1 = 3$, $x_2 = 0$ and can be written as $\mathbf{x} = \mathbf{By}$ where

$$\mathbf{B} = \begin{pmatrix} 1 & 0 \\ 0 & 0 \end{pmatrix}.$$

Although $\mathbf{A}^{-1}$ does not exist, the matrix $\mathbf{B}$ plays the role of an inverse in that its use leads to a solution to the equations. $\mathbf{B}$ is called a *generalized inverse* or *g-inverse* of $\mathbf{A}$, and is denoted by $\mathbf{A}^{-}$; it is sometimes called a pseudo-inverse or conditional inverse of $\mathbf{A}$. It is not unique since

$$\mathbf{A}^{-} = \begin{pmatrix} 1 & 0 \\ 2 & -1 \end{pmatrix}$$

also leads to the solution $x_1 = 3$, $x_2 = 0$, whereas

$$\mathbf{A}^{-} = \begin{pmatrix} 1/3 & 0 \\ 0 & 1/6 \end{pmatrix}$$

leads to a different solution, namely $x_1 = x_2 = 1$.

$\mathbf{A}^-$ is a g-inverse of $\mathbf{A}$ if and only if it satisfies

$$\mathbf{A}\mathbf{A}^-\mathbf{A} = \mathbf{A} \tag{A.15}$$

A solution to the consistent equations $\mathbf{A}\mathbf{x} = \mathbf{y}$ is then given by

$$\mathbf{x} = \mathbf{A}^-\mathbf{y} \tag{A.16}$$

It can be verified that the three g-inverses given in the above example satisfy (A.15).

The matrix $\mathbf{A}^+$ which satisfies the following four conditions:

$$\begin{aligned} \mathbf{A}\mathbf{A}^+\mathbf{A} &= \mathbf{A} \\ \mathbf{A}^+\mathbf{A}\mathbf{A}^+ &= \mathbf{A}^+ \\ (\mathbf{A}\mathbf{A}^+)' &= \mathbf{A}\mathbf{A}^+ \\ (\mathbf{A}^+\mathbf{A})' &= \mathbf{A}^+\mathbf{A} \end{aligned} \tag{A.17}$$

is called the *Moore–Penrose g-inverse* of $\mathbf{A}$; it can be shown to be an unique matrix.

If $\mathbf{A}$ is given in canonical form as

$$\mathbf{A} = \sum_{i=1}^{n} \lambda_i \mathbf{x}_i \mathbf{x}_i'$$

where $\lambda_1, \lambda_2, \ldots, \lambda_r$ $(r < n)$ are the non-zero eigenvalues, then:

(a)
$$\mathbf{A}^+ = \sum_{i=1}^{r} \lambda_i^{-1} \mathbf{x}_i \mathbf{x}_i' \tag{A.18}$$

is the Moore–Penrose g-inverse of $\mathbf{A}$;

(b)
$$\mathbf{A}^- = \sum_{i=1}^{r} \lambda_i^{-1} \mathbf{x}_i \mathbf{x}_i' + \sum_{i=r+1}^{n} \alpha_i \mathbf{x}_i \mathbf{x}_i' \tag{A.19}$$

is a g-inverse of $\mathbf{A}$ for a given set of constants $\alpha_{r+1}, \alpha_{r+2}, \ldots, \alpha_n$. If these constants are zero then $\mathbf{A}^- = \mathbf{A}^+$.

If $\mathbf{A}$ is a symmetric idempotent matrix then two useful g-inverses of $\mathbf{A}$ are the Moore–Penrose inverse $\mathbf{A}^+ = \mathbf{A}$ and the inverse $\mathbf{A}^- = \mathbf{I}$.

The following two results are frequently used in the book:

(a) $\mathbf{A}\mathbf{A}^-$ is an idempotent matrix with

$$\text{rank}(\mathbf{A}\mathbf{A}^-) = \text{trace}(\mathbf{A}\mathbf{A}^-) = \text{rank}(\mathbf{A}) \tag{A.20}$$

(b) If $(\mathbf{A}'\mathbf{A})^-$ is a g-inverse of $\mathbf{A}'\mathbf{A}$ then

$$\mathbf{A}(\mathbf{A}'\mathbf{A})^-\mathbf{A}'\mathbf{A} = \mathbf{A} \tag{A.21}$$

A.6 Kronecker products

Let $\mathbf{A} = ((a_{ij}))$ be a $p \times q$ matrix and $\mathbf{B}$ be an $r \times s$ matrix. The *Kronecker product* of $\mathbf{A}$ and $\mathbf{B}$, denoted by $\mathbf{A} \otimes \mathbf{B}$, is the $pr \times qs$ matrix defined as

$$\mathbf{A} \otimes \mathbf{B} = \begin{pmatrix} a_{11}\mathbf{B} & a_{12}\mathbf{B} & \cdots & a_{1q}\mathbf{B} \\ a_{21}\mathbf{B} & a_{22}\mathbf{B} & \cdots & a_{2q}\mathbf{B} \\ \vdots & \vdots & & \vdots \\ a_{p1}\mathbf{B} & a_{p2}\mathbf{B} & \cdots & a_{pq}\mathbf{B} \end{pmatrix}$$

Two special results are that $\mathbf{I}_n \otimes \mathbf{I}_m = \mathbf{I}_{nm}$ and $\mathbf{K}_n \otimes \mathbf{K}_m = \mathbf{K}_{nm}$.

If a is a scalar and matrices are conformable, then

$$\begin{aligned} (a\mathbf{A}) \otimes \mathbf{B} &= \mathbf{A} \otimes (a\mathbf{B}) = a(\mathbf{A} \otimes \mathbf{B}) \\ (\mathbf{A} \otimes \mathbf{B}) \otimes \mathbf{C} &= \mathbf{A} \otimes (\mathbf{B} \otimes \mathbf{C}) \\ (\mathbf{A} \otimes \mathbf{B})' &= \mathbf{A}' \otimes \mathbf{B}' \qquad \text{(A.22)} \\ (\mathbf{A} \otimes \mathbf{B})(\mathbf{C} \otimes \mathbf{D}) &= \mathbf{AC} \otimes \mathbf{BD} \\ (\mathbf{A} + \mathbf{B}) \otimes \mathbf{C} &= \mathbf{A} \otimes \mathbf{C} + \mathbf{B} \otimes \mathbf{C} \end{aligned}$$

Let $\mathbf{A}$ and $\mathbf{B}$ be square matrices of order n and m respectively; then

$$(\mathbf{A} \otimes \mathbf{B})^- = \mathbf{A}^- \otimes \mathbf{B}^- \qquad \text{(A.23)}$$

$$\text{trace}(\mathbf{A} \otimes \mathbf{B}) = \text{trace}(\mathbf{A}) \cdot \text{trace}(\mathbf{B}) \qquad \text{(A.24)}$$

If $\mathbf{x}_1$ is an eigenvector of $\mathbf{A}$ with eigenvalue λ_1, and $\mathbf{x}_2$ is an eigenvector of $\mathbf{B}$ with eigenvalue λ_2, then

$$(\mathbf{A} \otimes \mathbf{B})(\mathbf{x}_1 \otimes \mathbf{x}_2) = \lambda_1 \lambda_2 (\mathbf{x}_1 \otimes \mathbf{x}_2) \qquad \text{(A.25)}$$

i.e. $\mathbf{x}_1 \otimes \mathbf{x}_2$ is an eigenvector of $\mathbf{A} \otimes \mathbf{B}$ with eigenvalue $\lambda_1 \lambda_2$.

A.7 Circulant matrices

If $\mathbf{A} = ((a_{ij}))$ is an $n \times n$ matrix with $a_{ij} = a_{1m}$ where

$$m = \begin{cases} j - i + 1, & j \geqslant i \\ n - (j - i + 1), & j < i \end{cases}$$

then $\mathbf{A}$ is called a *circulant* matrix. Since the first row of a circulant matrix completely determines the matrix, the n elements in the first row of $\mathbf{A}$ will be denoted by $a_0, a_1, \ldots, a_{n-1}$. For example, the circulant matrix for $n = 4$ is

$$\mathbf{A} = \begin{pmatrix} a_0 & a_1 & a_2 & a_3 \\ a_3 & a_0 & a_1 & a_2 \\ a_2 & a_3 & a_0 & a_1 \\ a_1 & a_2 & a_3 & a_0 \end{pmatrix}$$

Let the $n \times n$ matrix $\mathbf{\Gamma}_h$ be a *basic* circulant matrix whose first row has 1 in the $(h+1)$th column and zero elsewhere. For instance, for $n = 4$,

$$\mathbf{\Gamma}_2 = \begin{pmatrix} 0 & 0 & 1 & 0 \\ 0 & 0 & 0 & 1 \\ 1 & 0 & 0 & 0 \\ 0 & 1 & 0 & 0 \end{pmatrix}$$

The general circulant matrix $\mathbf{A}$ can then be written as

$$\mathbf{A} = \sum_{h=0}^{n-1} a_h \mathbf{\Gamma}_h \tag{A.26}$$

The eigenvectors and eigenvalues of the circulant matrix $\mathbf{A}$ can be determined from those of $\mathbf{\Gamma}_1$, in view of (A.26) and the fact that

$$\mathbf{\Gamma}_h = \mathbf{\Gamma}_1^h \qquad (h = 0, 1, \ldots, n-1) \tag{A.27}$$

Eigenvectors of $\mathbf{\Gamma}_1$ are given by

$$\gamma_j = \frac{1}{\sqrt{n}} \begin{pmatrix} 1 \\ w^j \\ w^{2j} \\ \vdots \\ w^{(n-1)j} \end{pmatrix} \tag{A.28}$$

with corresponding eigenvalues

$$\lambda_{1j} = w^j$$

for $j = 0, 1, \ldots, n-1$, where w is given by

$$w = \exp(2\pi \mathrm{i}/n) = \cos(2\pi/n) + \mathrm{i}\sin(2\pi/n) \tag{A.29}$$

and where $\mathrm{i} = \sqrt{-1}$. This result follows from the fact that $w^n = 1$.

Using (A.7) and (A.27), eigenvectors of $\mathbf{\Gamma}_h$ are then given by (A.28) with eigenvalues

$$\lambda_{hj} = w^{jh}$$

for $h, j = 0, 1, \ldots, n-1$. Hence, eigenvectors of the general circulant matrix **A** are also given by (A.28) with eigenvalues

$$\lambda_j = \sum_{h=0}^{n-1} a_h w^{jh} \tag{A.30}$$

for $j = 0, 1, \ldots, n-1$. Note that the eigenvectors γ_j are independent of the elements of the matrix **A**.

If **A** is symmetric then $a_j = a_{n-j}$ $(j = 1, 2, \ldots, m)$ so that

$$\lambda_j = a_0 + \sum_{h=1}^{m} a_h[w^{jh} + w^{(n-j)h}] \tag{A.31}$$

where

$$m = \begin{cases} n/2, & n \text{ even} \\ (n-1)/2, & n \text{ odd} \end{cases}$$

Now since

$$\sin[2\pi(n-j)h/n] = -\sin(2\pi jh/n)$$

it follows that

$$w^{jh} + w^{(n-j)h} = \cos(2\pi jh/n) + \cos[2\pi(n-j)h/n]$$

so that the eigenvalues of a symmetric circulant matrix **A** are given by

$$\lambda_j = \sum_{h=0}^{n-1} a_h \cos(2\pi jh/n) \tag{A.32}$$

A spectral decomposition of **A** is given by

$$\mathbf{A} = \sum_{j=0}^{n-1} \lambda_j \gamma_j \bar{\gamma}_j' \tag{A.33}$$

where $\bar{\gamma}_j$ is the *conjugate* of γ_j given by replacing w in γ_j by its complex conjugate

$$\bar{w} = \exp(-2\pi i/n) = \cos(2\pi/n) - i\sin(2\pi/n)$$

Equation (A.33) can be verified by first showing that

$$\boldsymbol{\Gamma}_1 = \sum_{j=0}^{n-1} w^j \gamma_j \bar{\gamma}_j'$$

The (k, l)th element of $\gamma_j \bar{\gamma}_j'$ is $(1/n)w^{(k-l)j}$, and since for $k > l$

$$w^{(k-l)j} = w^{-[n-(k-l)]j}$$

it follows that $\gamma_j \bar{\gamma}_j'$ is a circulant matrix, which can be written as

$$\gamma_j \bar{\gamma}_j' = \frac{1}{n} \sum_{h=0}^{n-1} w^{-hj} \mathbf{\Gamma}_h \tag{A.34}$$

Hence

$$\sum_{j=0}^{n-1} w^j \gamma_j \bar{\gamma}_j' = \frac{1}{n} \sum_h \sum_j w^{-(h-1)j} \mathbf{\Gamma}_h = \mathbf{\Gamma}_1$$

since

$$\sum_j w^{-(h-1)j} = \begin{cases} n, & h = 1 \\ 0, & h \neq 1 \end{cases}$$

Then

$$\mathbf{A} = \sum_h a_h \mathbf{\Gamma}_1^h = \sum_j \{\sum_h a_h w^{jh}\} \gamma_j \bar{\gamma}_j'$$

which establishes (A.33). Note that

$$\gamma_j' \bar{\gamma}_k = \begin{cases} 1, & j = k \\ 0, & j \neq k \end{cases}$$

The Moore–Penrose inverse $\mathbf{A}^+$ of $\mathbf{A}$ is also a circulant matrix since it can be written as

$$\mathbf{A}^+ = \sum \lambda_j^{-1} \gamma_j \bar{\gamma}_j'$$

where the summation is over all r non-zero eigenvalues of $\mathbf{A}$, and where $r = \text{rank}(\mathbf{A})$. If $r = n$ then $\mathbf{A}^+ = \mathbf{A}^{-1}$. Now using (A.34),

$$\mathbf{A}^+ = \sum_h \theta_h \mathbf{\Gamma}_h$$

where

$$\theta_h = (1/n) \sum_j \lambda_j^{-1} w^{-hj}$$

If $\mathbf{A}$ is a symmetric matrix then, from (A.31), $\lambda_j = \lambda_{n-j}$ so that

$$\theta_h = (1/n) \sum_j \lambda_j^{-1} \cos(2\pi jh/n) \tag{A.35}$$

Let $\mathbf{\Gamma}_{h_k}$ be a basic circulant of order n_k and let the $n \times n$ matrix $\mathbf{\Gamma}_h$ be defined by

$$\mathbf{\Gamma}_h = \mathbf{\Gamma}_{h_1} \otimes \mathbf{\Gamma}_{h_2} \otimes \cdots \otimes \mathbf{\Gamma}_{h_m}$$

where $n = n_1 n_2 \ldots n_m$. Then following (A.26) a general *block circulant* matrix **A** is defined by

$$\mathbf{A} = \sum_{h_1=0}^{n_1-1} \sum_{h_2=0}^{n_2-1} \cdots \sum_{h_m=0}^{n_m-1} a_{h_1 h_2 \ldots h_m} \mathbf{\Gamma}_h \tag{A.36}$$

where $a_{h_1 h_2 \ldots h_m}$ is an element in the first row of **A**. It follows from (A.25) that eigenvectors of **A** are given by

$$\gamma_{j_1} \otimes \gamma_{j_2} \otimes \cdots \otimes \gamma_{j_m} \tag{A.37}$$

where γ_j is given by (A.28), with corresponding eigenvalues

$$\lambda_{j_1 j_2 \ldots j_m} = \sum_{h_1} \sum_{h_2} \cdots \sum_{h_m} a_{h_1 h_2 \ldots h_m} w^s$$

where

$$s = n \sum_{k=1}^{m} j_k h_k / n_k$$

for $j_k = 0, 1, \ldots, n_k - 1$.

If **A** is symmetric then

$$\lambda_{j_1 j_2 \ldots j_m} = \sum_{h_1} \cdots \sum_{h_m} a_{h_1 h_2 \ldots h_m} \cos\left[\sum_{k=1}^{m} (2\pi j_k h_k / n_k)\right] \tag{A.38}$$

References

Agrawal, H. (1966) Some methods of construction of designs for two-way elimination of heterogeneity. *J. Amer. Statist. Assoc.*, **61**, 1153–1171.

Agrawal, H.L. and Prasad, J. (1982a) Some methods of construction of balanced incomplete block designs with nested rows and columns. *Biometrika*, **69**, 481–483.

Agrawal, H.L. and Prasad, J. (1982b) Some methods of construction of GD-RC and rectangular-RC designs. *Austral. J. Statist.*, **24**, 191–200.

Bailey, R.A. (1977) Patterns of confounding in factorial designs. *Biometrika*, **64**, 597–603.

Bailey, R.A., Gilchrist, F.H.L. and Patterson, H.D. (1977) Identification of effects and confounding patterns in factorial designs. *Biometrika*, **64**, 347–354.

Basilevsky, A. (1983) *Applied Matrix Algebra in the Statistical Sciences.* North-Holland, New York.

Bose, R.C. (1947) Mathematical theory of the symmetrical factorial design. *Sankhyā*, **8**, 107–166.

Bose, R.C., Clatworthy, W.H. and Shrikhande, S.S. (1954) Tables of partially balanced designs with two associate classes. *North Carolina Agric. Expt. Station, Tech. Bull. 107.*

Bose, R. C. and Connor, W.S. (1952) Combinatorial properties of group divisible incomplete block designs. *Ann. Math. Statist.*, **23**, 367–383.

Bose, R.C. and Kishen, K. (1940) On the problem of confounding in general symmetrical factorial designs. *Sankhyā*, **5**, 21–36.

Bose, R.C. and Nair, K.R. (1939) Partially balanced incomplete block designs. *Sankhyā*, **4**, 337–372.

Bose, R.C. and Shimamoto, T. (1952) Classification and analysis of partially balanced incomplete block designs with two associate classes. *J. Amer. Statist. Assoc.*, **47**, 151–184.

Box, G.E.P., Hunter, W.G. and Hunter, J.S. (1978) *Statistics for Experimenters.* J. Wiley & Sons, New York.

Butz, L. (1982) *Connectivity in Multi-Factor Designs.* Heldermann Verlag, Berlin.

Ceranka, B. and Mejza, S. (1979) On the efficiency factor for a contrast of treatment parameters. *Biom. J.*, **21**, 99–102.

Clatworthy, W.H. (1973) *Tables of Two-Associate-Class Partially Balanced Designs*. Nat. Bureau of Stand., Appl. Math. Series 63.

Conniffe, D. and Stone, J. (1974) The efficiency factor of a class of incomplete block designs. *Biometrika*, **61**, 633–636.

Cotter, S.C. (1974) A general method of confounding for symmetrical factorial experiments. *J. Roy. Statist. Soc.*, B, **36**, 267–276.

Cox, D.R. (1958) *Planning of Experiments*. J. Wiley & Sons, New York.

David, H.A. (1967) Resolvable cyclic designs. *Sankhyā*, A, **29**, 191–198.

David, H.A. and Wolock, F.W. (1965) Cyclic designs. *Ann. Math. Statist.*, **36**, 1526–1534.

Davis, P.J. (1979) *Circulant Matrices*. J. Wiley & Sons, New York.

Dean, A.M. and John, J.A. (1975) Single replicate factorial experiments in generalized cyclic designs: II. Asymmetrical arrangements. *J. Roy. Statist. Soc.*, B, **37**, 72–76.

Draper, N.R. and Smith, H. (1981) *Applied Regression Analysis, Second Edition*, J. Wiley & Sons, New York.

Eccleston, J.A. and Hedayat, A. (1974) On the theory of connected designs: characterization and optimality. *Ann. Statist.*, **2**, 1238–1255.

Eccleston, J.A. and John, J.A. (1986) Recovery of row and column information in row-column designs with adjusted orthogonality. *J. Roy. Statist. Soc.*, B, **48**, in press.

Eccleston, J.A. and Kiefer, J. (1981) Relationships of optimality for individual factors of a design. *J. Statist. Planning and Inf.*, **5**, 213–219.

Eccleston, J.A. and Russell, K.G. (1975) Connectedness and orthogonality in multi-factor designs. *Biometrika*, **62**, 341–345.

Eccleston, J.A. and Russell, K.G. (1977) Adjusted orthogonality in non-orthogonal designs. *Biometrika*, **64**, 339–345.

Eccleston, J.A. and Russell, K.G. (1980) (M, S)-optimal row-column designs. *Commun. Statist. Theor. Meth.*, **A9**, 449–452.

Fisher, R.A. (1960) *The Design of Experiments*. Oliver and Boyd, Edinburgh.

Fisher, R.A. and Yates, F. (1963) *Statistical Tables for Biological, Agricultural and Medical Research*. Oliver and Boyd, Edinburgh.

Franklin, M.F. (1984) Constructing tables of minimum abberration p^{n-m} designs. *Technometrics*, **26**, 225–232.

Freeman, G.H. (1957) Some experimental designs of use in changing from one set of treatments to another. Part II: Existence of designs. *J. Roy. Statist. Soc.*, B, **19**, 163–165.

Freeman, G. H. (1976a) On the selection of designs for comparative experiments. *Biometrics*, **32**, 195–199.

Freeman, G.H. (1976b) A cyclic method of constructing regular group divisible incomplete block designs. *Biometrika*, **63**, 555–558.

Fries, A. and Hunter, W.G. (1980) Minimum aberration 2^{k-p} designs. *Technometrics*, **22**, 601–608.

Giovagnoli, A. (1977) On the construction of factorial designs using abelian group theory. *Rend. Sem. Mat. Univ. Padova*, **58**, 195–206.

Graybill, F.A. (1983) *Matrices with Applications in Statistics*. Wadsworth, Belmont.

Greenfield, A.A. (1976) Selection of defining contrasts in two-level experiments. *Appl. Statist.*, **25**, 64–67.

Greenfield, A.A. (1978) Selection of defining contrasts in two-level experiments – a modification. *Appl. Statist.*, **27**, 78.

Gupta, S.C. (1983) Some new methods for constructing block designs having orthogonal factorial structure. *J. Roy. Statist. Soc.*, B, **45**, 297–307.

Hall, W.B. and Jarrett, R.G. (1981) Nonresolvable incomplete block designs with few replicates. *Biometrika*, **68**, 617–627.

Harshbarger, B. (1949) Triple rectangular lattices. *Biometrics*, **5**, 1–13.

Ipinyomi, R.A. and John, J.A. (1985) Nested generalized cyclic row-column designs. *Biometrika*, **72**, 403–409.

James, A.T. and Wilkinson, G.N. (1971) Factorization of the residual operator and canonical decomposition of nonorthogonal factors in the analysis of variance. *Biometrika*, **58**, 279–294.

Jarrett, R.G. (1977) Bounds for the efficiency factor of block designs. *Biometrika*, **64**, 67–72.

Jarrett, R.G. (1983) Definitions and properties for *m*-concurrence designs. *J. Roy. Statist. Soc.*, B, **45**, 1–10.

Jarrett, R.G. and Hall, W.B. (1978) Generalized cyclic incomplete block designs. *Biometrika*, **65**, 397–401.

Jarrett, R.G. and Hall, W.B. (1982) Some designs considerations for variety trials. *Utilitas Mathematica*, **21B**, 153–168.

John, J.A. (1965) A note on the analysis of incomplete block experiments. *Biometrika*, **52**, 633–636.

John, J.A. (1969) A relationship between cyclic and PBIB designs. *Sankhyā*, **31**, 535–540.

John, J.A. (1981) Efficient cyclic designs. *J. Roy. Statist. Soc.*, B, **43**, 76–80.

John, J.A. and Dean, A.M. (1975) Single replicate factorial experiments in generalized cyclic designs: I. Symmetrical arrangements. *J. Roy. Statist. Soc.*, B, **37**, 63–71.

John, J.A. and Eccleston, J.A. (1986) Row–column α-designs. *Biometrika*, **73**, in press.

John, J.A. and Lewis, S.M. (1983) Factorial experiments in generalized cyclic row-column designs. *J. Roy.Statist. Soc.*, B, **45**, 245–251.

John, J.A. and Quenouille, M.H. (1977) *Experiments: Design and Analysis*. Griffin, London.

John, J.A. and Smith, T.M.F. (1974) Sum of squares in non-full rank general linear hypotheses. *J. Roy. Statist. Soc.*, B, **36**, 107–108.

John, J.A. and Turner, G. (1977) Some new group divisible designs. *J. Statist. Planning and Inf.*, **1**, 103–107.

John, J.A. and Williams, E.R. (1982) Conjectures for optimal block designs. *J. Roy. Statist. Soc.*, B, **44**, 221–225.

John, J.A., Wolock, F.W. and David, H.A. (1972) *Cyclic Designs*. Nat. Bureau of Stand., Appl. Math. Series 62.

John, P.W.M. (1971) *Statistical Design and Analysis of Experiments*. Macmillan, New York.

Jones, B. (1976) An algorithm for deriving optimal block designs. *Technometrics*, **18**, 451–458.

Jones, B. and Eccleston, J.A. (1980) Exchange and interchange procedures to search for optimal designs. *J. Roy. Statist. Soc.*, B, **42**, 238–243.

Kempthorne, O. (1947) A simple approach to confounding and fractional replication in factorial experiments. *Biometrika*, **34**, 255–272.

Kshirsagar, A.M. (1957) On balancing in designs in which heterogeneity is eliminated in two directions. *Calcutta Statist. Assoc. Bull.*, **7**, 161–166.

Kshirsagar, A.M. (1966) Balanced factorial designs. *J. Roy. Statist. Soc.*, B, **28**, 559–567.

Kurkjian, B. and Zelen, M. (1963) Applications of the calculus for factorial arrangements: I. Block and direct product designs. *Biometrika*, **50**, 63–73.

Lamacraft, R.R. and Hall, W.B. (1982) Tables of cyclic incomplete block designs: $r = k$. *Austral. J. Statist.*, **24**, 350–360.

Lewis, S.M. (1982) Generators for asymmetrical factorial experiments. *J. Statist. Planning and Inf.*, **6**, 59–64.

Lewis, S.M. and Dean, A.M. (1985) A note on efficiency-consistent designs. *J. Roy. Statist. Soc.*, B, **47**, 261–262.

McLean, R.A. and Anderson, V.L. (1984) *Applied Factorial and Fractional Designs*. Dekker, New York.

Mukerjee, R. (1979) Inter-effect-orthogonality in factorial experiments. *Calcutta Statist. Assoc. Bull.*, **28**, 83–108.

Mukerjee, R. (1981) Construction of effect-wise orthogonal factorial designs. *J. Statist. Planning and Inf.*, **5**, 221–229.

National Bureau of Standards (1957) Fractional factorial experiment designs for factors at two levels. *Appl Math. Series* 48.

National Bureau of Standards (1959) Fractional factorial experiment designs for factors at three levels. *Appl. Math. Series* 54.

National Bureau of Standards (1961) Fractional factorial designs for experiments with factors at two or three levels. *Appl. Math. Series* 58.

Nelder, J.A. (1965) The analysis of randomized experiments with orthogonal block structure: I. Block structure and the null analysis of variance; II. Treatment structure and the general analysis of variance. *Proc. Roy. Soc.*, A, **283**, 147–178.

Nelder, J.A. (1968) The combination of information in generally balanced designs. *J. Roy. Statist. Soc.*, B, **30**, 303–311.

Paterson, L.J. (1983) Circuits and efficiency in incomplete block designs. *Biometrika*, **70**, 215–225.

Paterson, L.J. and Patterson, H.D. (1983) An algorithm for constructing α-lattice designs. *Ars Combinatoria*, **16A**, 87–98.

Patterson, H.D. (1965) The factorial combination of treatments in rotation experiments. *J. Agric. Sci.*, **65**, 171–182.

Patterson, H.D. (1976) Generation of factorial designs. *J. Roy. Statist. Soc.*, B, **38**, 175–179.

Patterson, H.D. and Bailey, R.A. (1978) Design keys for factorial experiments. *Appl. Statist.*, **27**, 335–343.

Patterson, H.D. and Silvey, V. (1980) Statutory and recommended list trials of crop varieties in the United Kingdom. *J. Roy. Statist. Soc.*, A, **143**, 219–252.

Patterson, H.D. and Thompson, R. (1971) Recovery of inter-block information when block sizes are unequal. *Biometrika*, **58**, 545–554.

Patterson, H.D. and Williams, E.R. (1976a) A new class of resolvable incomplete block designs. *Biometrika*, **63**, 83–92.

Patterson, H.D. and Williams, E.R. (1976b) Some theoretical results on general block designs. *Proc. 5th British Combin. Conf. Congressus Numerantium* **XV**, 489–496. Utilitas Mathematica, Winnipeg.

Patterson, H.D., Williams, E.R. and Hunter, E.A. (1978) Block designs for variety trials. *J. Agric. Sci.*, **90**, 395–400.

Pearce, S.C. (1963) The use and classification of non-orthogonal designs. *J. Roy. Statist. Soc.*, A, **126**, 353–377.

Pearce, S.C. (1974) Optimality of design in plot experiments. *Proc. 8th Int. Biometric Conf.*, Constanta, Romania.

Pearce, S.C. (1975) Row-and-column designs. *Appl. Statist.*, **24**, 60–74.

Pearce, S.C., Calinski, T. and Marshall, T.F. de C. (1974) The basic contrasts of an experimental design with special reference to the analysis of data. *Biometrika*, **61**, 449–460.

Preece, D.A. (1967) Nested balance incomplete block designs. *Biometrika*, **54**, 479–486.

Raghavarao, D. (1971) *Constructions and Combinatorial Problems in Design of Experiments*. J. Wiley & Sons, New York.

Raghavarao, D. and Federer, W.T. (1975) On connectedness in two-way elimination of heterogeneity designs. *Ann. Statist.*, **3**, 730–735.

Raghavarao, D. and Shah, K.R. (1980) A class of D_0 designs for two-way elimination of heterogeneity. *Commun. Statist. Theor. Meth.*, **A9**, 75–80.

Raktoe, B.L. (1969) Combining elements from distinct finite fields in mixed factorials. *Ann. Math. Statist.*, **40**, 498–504.

Rao, C.R. (1946) Confounded factorial designs in quasi-Latin squares. *Sankhyā*, **7**, 295–304.

Rice, C.G. and Izumi, K. (1984) Annoyance due to combinations of noises. *Proc. Inst. Acoustics*, 287–294.

Roy, J. and Shah, K.R. (1961) Analysis of two-way designs. *Sankhyā*, A, **23**, 129–144.

Russell, K.G. (1976) The connectedness and optimality of a class of row-column designs. *Commun. Statist. Theor. Meth.*, **A5**, 1479–1488.

Russell, K.G. (1980) Further results on the connectedness and optimality of designs of type O:XB. *Commun. Statist. Theor. Meth.*, **A9**, 439–447.

Searle, S.R. (1966) *Matrix Algebra for the Biological Sciences*. J. Wiley & Sons, New York.

Searle, S.R. (1971) *Linear Models*. J. Wiley & Sons, New York.

Shah, B.V. (1959) A generalization of partially balanced incomplete block designs. *Ann. Math. Statist.*, **30**, 1041–1050.

Shah, K.R. (1960) Optimality criteria for incomplete block designs. *Ann. Math. Statist.*, **31**, 791–794.

Shah, K.R. (1962) On estimation of intergroup variance in one- and two-way designs. *Sankhyā*, A, **24**, 281–286.

Shah, K.R. (1977) Analysis of designs with two-way elimination of heterogeneity. *J. Statist. Planning and Inf.*, **1**, 207–216.

Shrikhande, S.S. (1951) Designs for two-way elimination of heterogeneity. *Ann. Math. Statist.*, **22**, 235–247.

Sihota, S.S. and Banerjee, K.S. (1981) On the algebraic structures in the construction of confounding plans in mixed factorial designs on the lines of White and Hultquist. *J. Amer. Statist. Assoc.*, **76**, 996–1001.

Singh, M. and Dey, A. (1979) Block designs with nested rows and columns. *Biometrika*, **66**, 321–326.

Street, D.J. (1981) Graeco-Latin and nested row and column designs. *Combinatorial Mathematics*, **8**, 304–313. Lecture Notes in Mathematics 884. Ed. K.L. McAvaney. Springer-Verlag, Berlin.

Voss, D.T. and Dean, A.M. (1987) A comparison of classes of single replicate factorial designs. *Ann. Statist.*, **15**, in press.

Wheeler, N.C. and Kshirsagar, A.M. (1981) Uniformly better estimators with applications in two-way designs. *Commun. Statist. Theor. Meth.*, **A10**, 699–711.

White, D. and Hultquist, R.A. (1965) Construction of confounding plans for mixed factorial designs. *Ann. Math. Statist.*, **36**, 1256–1271.

Williams, E.R. (1975a) A new class of resolvable block designs. Ph.D. Thesis, Univ. of Edinburgh.

Williams, E.R. (1975b) Efficiency-balanced designs. *Biometrika*, **62**, 686–689.

Williams, E.R. (1976) Resolvable paired-comparison designs. *J. Roy. Statist. Soc.*, B, **38**, 171–174.

Williams, E.R. and Patterson, H.D. (1977) Upper bounds for efficiency factors in block designs. *Austral. J. Statist.*, **19**, 194–201.

Williams, E.R., Patterson, H.D. and John, J.A. (1976) Resolvable designs with two replications. *J. Roy. Statist. Soc.*, B, **38**, 296–301.

Williams, E.R., Patterson, H.D. and John, J.A. (1977) Efficient two-replicate resolvable designs. *Biometrics*, **33**, 713–717.

Worthley, R. and Banerjee, K.S. (1974) A general approach to confounding plans in mixed factorial experiments when the number of levels of a factor is any positive integer. *Ann. Statist.*, **2**, 579–585.

Yates, F. (1936a) Incomplete randomized designs. *Ann. Eugenics*, **7**, 121–140.

Yates, F. (1936b) A new method of arranging variety trials involving a large number of varieties. *J. Agric. Sci.*, **26**, 424–455.

Yates, F. (1937) A further note on the arrangement of variety trials: Quasi-Latin squares. *Ann. Eugenics*, **7**, 319–331.

Yates, F. (1939) The recovery of inter-block information in variety trials arranged in three-dimensional lattices. *Ann. Eugenics*, **9**, 136–156.

Yates, F. (1940a) Lattice squares. *J. Agric. Sci.*, **30**, 672–687.

Yates, F. (1940b) The recovery of inter-block information in balanced incomplete block designs. *Ann. Eugenics*, **10**, 317–325.

Author index

Subject index